Bifurcation Analysis

Principles, Applications and Synthesis

Bifurcation Analysis

Principles, Applications and Synthesis

edited by

M. Hazewinkel

R. Jurkovich

and

J. H. P. Paelinck

Steering Committee on Interdisciplinary Studies,
Erasmus University Rotterdam, The Netherlands

D. REIDEL PUBLISHING COMPANY

A MEMBER OF THE KLUWER ACADEMIC PUBLISHERS GROUP

DORDRECHT / BOSTON / LANCASTER

Library of Congress Cataloging in Publication Data

Main entry under title:

Bifurcation analysis.

    Includes index.
      1. Differential equations, Nonlinear–Numerical solutions.
2. Bifurcation theory. I. Hazewinkel, Michiel. II. Jurkovich, Ray.
III. Paelinck, Jean H. P. IV. Erasmus Universiteit Rotterdam.
Steering Committee on Interdisciplinary Studies.
QA372.B53 1984   515.3′55   84–18254
ISBN 90-277-1446-0

Published by D. Reidel Publishing Company,
P.O. Box 17, 3300 AA Dordrecht, Holland.

Sold and distributed in the U.S.A. and Canada
by Kluwer Academic Publishers,
190 Old Derby Street, Hingham, MA 02043, U.S.A.

In all other countries, sold and distributed
by Kluwer Academic Publishers Group,
P.O. Box 322, 3300 AH Dordrecht, Holland.

TABLE OF CONTENTS

C. Synthesis

PREFACE

Bifurcation theory has made a very fast upswing in the last
fifteen years.

Roughly speaking it generalises to dynamic systems the possibility of multiple solutions, a possibility already recognised in
static systems - physical, chemical, social - when operating far
from their equilibrium states.

It so happened that quite a few staff members of the Erasmus
University Rotterdam were thinking along those lines about certain
aspects of their disciplines. To have a number of specialists and
potential "fans" convene to discuss various aspects of bifurcational thinking, seemed a natural development. The resulting papers
were judged to be of interest to a larger public, and as such are
logically regrouped in this volume, one in a series of studies
resulting from the activities of the Steering Committee on
Interdisciplinary Studies of the Erasmus University, Rotterdam.

Although the volume is perhaps multidisciplinary rather
than interdisciplinary - the interdisciplinary aspect being only
"latent" -, as a "soft" interdisciplinary exercise (the application
of formal structures of one discipline to another) it has a right
to interdisciplinary existence!

This book could not have been published without a generous
grant of the University Foundation of the Erasmus University
Rotterdam, which allowed the conference to be held and the
resulting papers to be published; that generosity is gratefully
acknowledged.

Work based on the ideas exposed in this volume is going on
in the various centres to which the authors belong; the editors
hope that these ideas will be stimulating to readers in various
fields of scientific research.

M. Hazewinkel
R. Jurkovich
J. H. P. Paelinck

A. Principles

SELF-ORGANISATION IN NONEQUILIBRIUM SYSTEMS: TOWARDS A DYNAMICS OF COMPLEXITY

I. Prigogine and G. Nicolis

Faculté des Sciences Université Libre de Bruxelles and Centre for studies in Statistical Mechanics, University of Texas

1. INTRODUCTION

Complex behaviour made its appearance in the physical sciences in a modest, low-key fashion. For a long time, in the mind of most physicists and chemists, complexity was associated with biological order and its multiple manifestations, for example at the level of evolution, embryogenesis and population dynamics. Physical sciences on the other side were aiming at a description of nature in terms of laws of universal validity. And to this end they were utilising simple models to which, hopefully, the description of more complicated systems could be reduced. This feeling has been repreatedly expressed by some of the greatest scientists of our century. Thus, for Einstein "the physicist must content himself with describing the most simple events which can be brought within the domain of our experience; all events of a more complex order are beyond the power of the human intellect to reconstruct with the subtle accuracy and logical perfection which the theoretical physicist demands .... The general laws on which the structure of theoretical physics is based claim to be valid for any natural phenomenon whatsoever. With them, it ought to be possible to arrive at the description, that is to say, the theory, of every natural process including life, by means of pure deduction, if that process of deduction were not far beyond the capacity of the human intellect".

True, from time to time some fissures were appearing, threatening the status of this magnificent edifice: the discovery of the erratic behaviour of turbulence flows by

3

M. Hazewinkel et al. (eds.), Bifurcation Analysis, 3–12.
© 1985 by D. Reidel Publishing Company.

Reynolds, the periodic precipitation phenomena known as Liesegang rings, or the oscillatory behaviour of certain chemical
oscilliations. The usual reaction to the challenge, however,
was either to regard such phenomena as curiosities and even
artifices, or to convince oneself that they did not really
belong to the realm of physical sciences being rather parts of
engineering, materials science, biology or geology.

The last twenty years have witnessed a radical departure
from this monolithic point of view. We are still in the middle
of this reconceptualisation, but we can already perceive a new
attitude in the description of nature. In brief, this new attitude corresponds to a "rediscovery of time" at all levels of
description in science. The two great revolutions in physics of
this century are quantum mechanics and relativity. Both started
as corrections to classical mechanics, made necessary once the
role of the universal constants, c (the velocity of light) and
h (Planck's constant), was discovered. Today both have taken an
unexpected "temporal" turn: quantum mechanics deals in its most
interesting part with the description of unstable particles and
their mutual transformations. Similarly, relativity started as
a geometrical theory, but today this theory is mainly concerned
with the thermal history of the early universe. More striking
from our point of view is the fact that radical changes are
taking place in fields reputed to be classical and well established like dynamics or chemistry. Simple, high-school examples
of mechanical systems are now known to present unexpectedly
complex behaviour. A periodically forced pendulum on the
borderline between vibration and rotation gives rise to a rich
variety of motions, including the possibility of random
turbulent-like excursions from its equilibrium point. Chemical
reactions, usually thought to reach rapidly a stationary state
of uniform composition, can generate a multitude of self-
organisation phenomena in the form of spatial structures,
temporal rhythms, or propagating wave-fronts.

As a result of these discoveries the interest towards
macroscopic physics, the physics dealing with phenomena in our
scale, is increasing enormously. One of the most striking
aspects of this revival is the realisation that the distinction
between "simple" and "complex", between "disorder" and "order",
is much narrower than usually thought: complexity is no longer
confined to biology, but is invading physical sciences as well.
The description of the phenomena arising in this context is the
subject of a new discipline, which may be called tentatively
the dynamics of complex systems.

The aim of this article is to describe some of the avenues
along which progress is likely to occur in this new discipline.
But let us first sketch briefly its present status.

2. THE DYNAMICS OF COMPLEX SYSTEMS

A first facet of the dynamics of complex systems corresponds to thermodynamics. It is indeed necessary to characterise
the onset of order and complex behaviour at the macroscopic
level from the standpoint of the theory of irreversible processes. This approach started in the 1940's, after Onsager's
pioneering work in the theory of fluctuations, when experimental evidence was still scarce, and culminated in the late
1960's in the discovery of the concept of dissipative structure. The existence of such structures reflects the fact that
irreversibility, dissipation, and distance from equilibrium can
act as source of order. No simple extrapolation of the equilibrium like behaviour can thus be expected far from equilibrium:
new dynamical states of matter like chemical oscillations or
regular convection cells, which would be inconceivable under
equilibrium conditions, become possible in this range.
Realising this is a surprise comparable, say, to the one that
would experience intelligent beings accustomed to live in a
high temperature environment at the view of a beautiful
crystal!

More specifically irreversibility and dissipation confer
to physical systems the property of asymptotic stability, that
is to say, the ability to damp the effect of disturbances
acting on them. Thanks to this property, which is absent in
conservative systems like those one deals with in mechanics,
sharp and reproducible configurations of matter can be sustained. On the other hand, the distance from equilibrium allows
the system to fully "reveal" the potentialities hidden in the
feedbacks and other nonlinearity of the kinetics, and thus to
undergo transitions to dynamical behaviour.

Despite the conceptual importance of the above considerations, it is clear that at some stage of the analysis additional information coming from the mathematical structure of the
evolution equations is necessary. In this way we are led to the
fascinating, and still largely unexplored world of dynamical
systems. As it turns out the onset of complex behaviour can be
associated with a mathematical concept of a great generality,
the bifurcation of new branches of solutions from some
"reference" state. Moreover, bifurcation is by no means a
unique event. Rather, it is the beginning of a complex sequence
of transitions which may lead to symmetry-breaking and space-
time order, to a high multiplicity of solutions and the con-
comitant ability of regulation and switching, but also to an
erratic time evolution commonly referred to as "chaos" (cf.
Figure 1). This marks in fact the onset of what was long known
as turbulence in fluid dynamics. The new fascinating aspect is,
however, that such behaviour is no more confined to a single
field but, rather, invades such different disciplines as
chemistry, optics, or metallurgy.

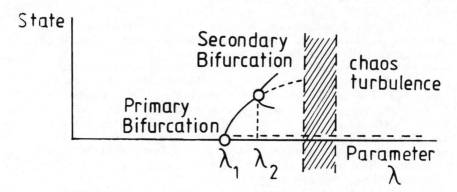

Figure 1.

It is striking to see such radically different kinds of
behaviour embedded in a single description. On the other side
we have the first bifurcation branches, which reflect the
appearance of strict regulations and controls ensuring order.
It is tempting to associate such regimes with biological order.
And on the other side we have turbulent chaos, reflecting the
collapse of an order encompassing the system as a whole. One
can hardly avoid the feeling that such states should provide
the physico-chemical basis for understanding pathological behaviour and disease. In short, we are witnessing both an incredible specificity and diversity of nonequilibrium states as
well as a deep unity in the basic concepts involved in their
description.

The generality of the phenomenon of bifurcation in non-
linear dynamical systems leads us quite naturally to a third
facet of complex systems, namely the stochastic analysis of
fluctuations. Indeed, bifurcation is basically a "decision
making" process. But because of the multiplicity of choices at
the "decision" point a selection mechanism is necessarily
involved. Now all physical systems possess such a mechanism, in
the form of random fluctuations that are generated spontaneously around the deterministic evolution. Thanks to them the state
space is continuously scanned, and the evolution is directed
toward the appropriate outcome. This adds therefore a ubiqui-
tous historical dimension in the description of the system.

The fourth, but by no means the least important facet of
complex systems is, naturally, experiment. After a laborious
induction period we have witnessed an explosion of experimental
results in the last ten years. New sophisticated techniques,
many of them based on laser spectroscopy, have increased both
the precision and range of space and time scales that could be
explored. Moreover, a new set of quantities, related to the
distance from equilibrium, have imposed themselves as key

control parameters in the experimental technology. The resi-
dence time of initial products in a chemical reactor is a
typical example.

The message resulting from these experimental investigations is, once again, that different systems present similar
sequences of transitions; and simple behaviour coexists side-
by-side with a variety of complex phenomena ranging from sharp
periodicities to chaos. It suffices to modify some of the
externally controlled parameters to see all these behaviours
come into existence and parade in the laboratory! Moreover,
many of these transitions occur in situations of great interest
in applied sciences like, for instance, heterogeneous cata-
lysis. In this respect it is worth noting that even reactions
having a very simple kinetics can give rise to complex dynamics
if they take place in a sufficiently complex environment such
as the surface of a catalyst.

Several characteristic features of the present state of
dynamics of complex systems become obvious from the above brief
presentation. Both macroscopic theories and theories dealing
with fluctuations are needed to understand complexity, as both
"chance" and "necessity" are cooperating in the emergence of
new types of behaviour. The systematic use of probabilistic
concepts, longtime confined to quantum mechanics, is therefore
making its entrance in the study of macroscopic phenomena. More
generally after a period of parallel practically unrelated
developments, the thermodynamic and statistical theory of
irreversible processes, the theory of dynamical systems and the
new experimental techniques become now parts of a vast interdisciplinary effort in which intense cross-fertilization
between different ideas and methods begins to take place.
Complex behaviour no longer appears to be a singularity in the
otherwise uneventful history of a physical system. Rather, it
is realised that it is deeply rooted into physics, that it may
emerge and disappear repeatedly as the conditions vary, and
even that it can coexist with the more familiar simple static
behaviour.

3. PERSPECTIVES

After this brief summary let us look forward to the future
developments in the field of complex systems. It seems to us
that the most exciting discoveries are still ahead of us.

From the standpoint of nonlinear dynamics and bifurcation
phenomena it would not be unfair to qualify our present knowledge as some sort of "taxonomy", the art of systematics prevailing in botany and zoology before the advent of molecular
biology! This is largely due to the global character of most of
these phenomena, as a result of which the problems remain
highly nonlinear and thus intractable. True, from time to time

some beautiful regularities emerge and inspire unexpected progress: the discovery of universality in bifurcation cascades
leading to a doubling of the period in certain classes of discrete time mappings is a striking example. Still, the vast
majority of transition phenomena remains poorly characterised.
How these global phenomena can become amenable to a local
analysis utilising perturbation theory is a major challenge of
the coming years. The experimental discovery that different
types of complex behaviour are possible in narrow ranges of
parameter values gives some hope that such a local view of
complexity might become a reality.

Mastering the mechanisms of bifurcation cascades will
undoubtedly allow us to have a new look at the particularly
fascinating problem of biological order. It is already clear
that the fact that matter may possess properties usually
ascribed to life like sensitivity, choice, history, narrows the
gap between "life" and "non-life". But can one go one step
further toward an understanding of generation of life? Manfred
Eigen made an important contribution in this direction by
showing that populations of biopolymers endowed with the
ability of self-replication and autocatalysis can evolve to
states of increasing complexity, which may be thought of as
precursors of the genetic code. One major question remains open
however, namely, how can self-replication and autocatalysis
come about. Prebiotic chemistry gives some interesting clues by
showing that, at the molecular level, small amino-acid and
nucleotide chains are capable of cooperative behaviour. It will
be a challenge of the coming years to understand how out of
such elementary subunits a supermolecular organisation capable
of utilising autonomously the resources of the environment,
like a bioenergetic pathway, can emerge. It is difficult to
avoid the feeling that bifurcations and self-organisation in
nonequilibrium conditions should play a key role in this
venture.

As we repeatedly stressed the occurrence of bifurcations
reveals the new potentialities of matter under nonequilibrium
conditions. On the one side we know that matter is capable of
coherent behaviour at the macroscopic scale. But on the other
side, because of the multiplicity of regimes that can be
observed we realise that statistical considerations should be
present in physical sciences, including branches reputed to be
eminently deterministic like chemistry. Let us mention a new
and still practically unexplored aspect of this duality,
suggested by recent studies of fluctuations in nonequilibrium
systems. It appears that in systems characterised by nonlinear
kinetics internal differentiation may take place not only in
the space of the parameters (as it usually happens in bifurcation) but also in time, as the system follows the course of
its otherwise deterministic evolution. It suffices for this
that the rate of change of the variables switches from low to

high values at some characteristic "ignition" moments. What is
happening then is that, because of fluctuations, different
parts of the system perceive different "ignition" times and
consequently some of them evolve ahead of others. This phenomenon of "bifurcations unfolding in time" shows therefore the
enormous plasticity of matter when the evolution occurs under
highly nonequilibrium conditions. It should play an important
role in the understanding of phenomena involving sudden tran-
sients, like combusiton of explosions.

   So far we have been concerned primarily with the mechanisms generating complex behaviour in physico-chemical
systems. It is clear, however, that modelling complexity and
evolution as a sequence of bifurcations in which fluctuations
provide the mechanism of selection is a picture extending far
beyond the range of physics and chemistry. Concepts developed
in connection with the dynamics of complex systems are thus
likely to be in the forefront in the extensive transfer of
knowledge that is going to take place in the forthcoming years
between physics and chemistry on the one side and environmental, biological or human sciences on the other.

   Let us illustrate on a representative example how we view
this problem of transfer of knowledge. Consider the problem of
climatic variability. In the first half of the present century
mankind has experienced a particularly mild and predictable
climate. It is now established, however, that this was an
exception in climate's long turbulent history. On all known
scales, ranging from one season to tens of thousands of years,
climate has experienced massive changes. In an overpopulated
and energy-thirsty society like ours the possibility that such
changes can indeed take place constitutes a major question to
be faced, if not to be mastered. Whence the recent interest in
modelling climatic change.

   The first problem arising is whether climatic change is a
phenomenon intrinsic to the earth-atmosphere-cryosphere-bio-
sphere system, or whether it is an externally driven process.
To be specific take the example of glaciations of the quaternary era, which are certainly the most dramatic climatic epi-
sodes of the last two million years. These phenomena occurred
with a relatively well-defined average periodicity of the order
of 100,000 years. This happens to be precisely the periodicity
of variation of the eccentricity of the earth's orbit, as a
result of which the solar energy received on the earth's upper
atmosphere slowly varies with a relative amplitude of about
0.001. We arrive therefore at an apparent paradox: on the one
side, the correlation between the two periodicities is impressive and on the other side, the amplitude of the external
forcing acting on the climatic system is too small to trigger a
major change.

   It is here that the utilisation of concepts stemming from
the dynamics of complex systems permits us to arrive at a

coherent picture. Climate dynamics is modelled by a set of
highly nonlinear balance equations, which as a rule admit
multiple solutions. One of them corresponds to present-day
climate, and others to glacial-like climates. Both types of
climate are possible and compatible with all known constraints,
in other words, they both constitute stable solutions of the
climatic equations. Which one will be preferred? We know from
the theory of complex systems that fluctuations provide the
selection mechanism. So let us study the influence of fluctuations under the action of a periodic forcing representing the
100,000 year cycle of orbital variations. The result is very
surprising: because of the fluctuations the weak orbital signal
is amplified drastically and entrains the whole system into a
100,000 year cycle corresponding to massive climatic changes
reminiscent of quaternary glaciations. In this way we arrive at
a synthesis between external and interally generated mechanisms
of climatic change.

4. CONCLUDING REMARKS

We have seen that the study of matter under nonequilibrium
conditions introduces new concepts and, in particular, shows
the constructive role of irreversible phenomena in nature. We
are convinced that the modifications that we have to impose in
our description of matter as a result of these discoveries will
not be restricted to macroscopic physics. Important repercuss-
ions have to be expected also in the extreme scales of natural
phenomena, particle physics and cosmology. As already mentioned
in this article, our view of the physical world is shifting
from a mechanistic, reversible description to a thermodynamic
one in which notions such as entropy, instability, evolution,
will play an ever increasing role. Signposts in this direction
are the theory of black holes in cosmology and the gauge
theories of unified interactions in elementary particle
physics.

BIBLIOGRAPHICAL NOTE

Because of the shortness of this text we could not give an extensive coverage of the literature. However, we would like to list a few textbooks, monographs or papers on some of the main directions of research in the area of complex systems.

(i) Thermodynamic, bifurcation, and stochastic analyses of irreversible processes.

Glansdorff, P. and Prigogine, I.: 1971, Thermodynamic Theory of Structure, Stability and Fluctuations, Wiley-Interscience, London.

Prigogine, I.: 1980, From being to becoming, W.H. Freeman and Company, San Francisco.

Nicolis, G. and Prigogine, I.: 1977, Self-Organization in Nonequilibrium Systems, Wiley-Interscience, New York.

Haken, H.: 1977, Synergetics, Springer, Berlin.

(ii) Bifurcation theory

Sattinger, D.: 1973, Topics in Stability and Bifurcation Theory, Springer, Berlin.

Arnold, V.: 1980, Chapitres complémentaires de la théorie des équations différentielles ordinaires, Mir, Moscow.

(iii) Routes to chaos

Feigenbaum, M.: 1978, Quantitative universality for a class of nonlinear transformations, J. Stat. Phys. 19, 25; 1979, 21, 669.

Pacault, A. and Vidal, C. (eds.): 1981, Nonlinear Phenomena in Chemical Dynamics, Springer, Berlin.

(iv) Fluctuations in nonequilibrium systems

Nicolis, G. and Prigogine, I.: 1971, Fluctuations in Nonequilibrium Systems, Proc. Nat. Sci. (U.S.A.) 68, 2102.

Nicolis, G., Dewel, G. and Turner J.W. (eds.): 1981, Order and Fluctuations in Equilibrium and Nonequilibrium Statistical Mechanics, Wiley, New York.

(v) Experimental aspects

Pacault, A. and Vidal, C.: Loc. cit.

Swinney, H. and Gollub, J.: 1981, Hydrodynamic Instabilities and the Transition to turbulence, Springer, Berlin.

(vi) Biological evolution

Eigen, M. and Schuster, P.:1979, The Hypercycle, Springer, Berlin.

Prigogine, I., Nicolis G. and Babloyantz, A.: 1972, Thermodynamics of Evolution, Physics Today, 25, nos. 11 and 12.

(vii) Climate dynamics

Nicolis, C.: 1982, Stochastis Aspects of Climatis Transitions: Response to a Periodic forcing, Tellus no. 34, 1

Benzi, R., Parisi, G., Sutera, A. and Vulpiani, A.: 1982, Stochastic resonance in climate change, Tellus no. 34, 10.

BIFURCATION PHENOMENA. A SHORT INTRODUCTORY TUTORIAL WITH EXAM-
PLES

Michiel Hazewinkel

Centre for Mathematics and Computer Science, Amsterdam
& Econometric Institute, Erasmus University Rotterdam

1. INTRODUCTION

Many problems in the physical and the social sciences can
be described (modelled) by equations or inequalities of one
kind or another. E.g. simple polynomial equations such as

$x^3 - 2x^2 + 3x - 4 = 0$ or a difference equation $x(t+1) =$

$2y(t)+x(t)$, $y(t+1) = 2x(t)-y(t)$, or a differential equation

$\dot{x}(t) = - x^2(t) + \sin t$, or much more complicated equations

such as integro-differential equations, etcetera. In such a
case a large part of solving the problem consists of solving
the equation(s) and describing various properties of the nature
of the solution (such as stability). Almost always such
equations contain a number of parameters whose values are
determined by the particular phenomenon being modelled. These
are then usually not exactly known and may even change in time
either in a natural way or because they are in the nature of
control variables which can be adjusted to achieve certain
goals.
    Thus it becomes natural and important to study families of
equations, e.g. $x(t+1) = f(x,\lambda)$ or $g(x,\lambda) = 0$ depending on a
parameter λ and to study how the set of solutions of such an
equation varies (in nature) as the parameters vary. This is the
topic of bifurcation theory. Roughly speaking $\lambda_0 \in \Lambda$ (= parameter space) is a bifurcation point if the nature of the set of
solutions of the family of equations changes at point λ_0.
    This chapter tries mainly by means of a few examples from

M. Hazewinkel et al. (eds.), Bifurcation Analysis, 13–30.
© 1985 by D. Reidel Publishing Company.

physics and economics to give a first idea of what bifurcation
theory is about.

2. EXAMPLE. BALL IN A HILLY LANDSCAPE

Consider a one dimensional landscape (depending on a parameter [1]) given by the (potential energy) expression

$$E(\lambda,x) = \tfrac{1}{4}x^4 - x^2 - (2\lambda - 4) x + 9 \qquad (2.1)$$

E.g. if $\lambda = 2$, then $E(\lambda,x)$ looks as follows

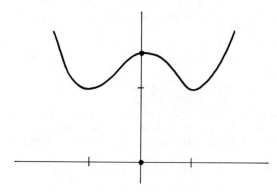

Figure 1.

We are interested in finding the equilibrium positions of a
ball in this landscape. If $\lambda = 2$ these equilibrium positions
are clearly $x = -\sqrt{2}$, $x = 0$ and $x = \sqrt{2}$.

In general these equilibrium positions are given by

$$E_x(\lambda,x) = x^3 - 2x - (2\lambda - 4) = 0 \qquad (2.2)$$

Depending on λ this equation has one, two or three solutions.
To be precise it has one solution for $\lambda < 2 - \frac{2}{9}\sqrt{6}$ and
$\lambda > 2 + \frac{2}{9}\sqrt{6}$, it has two solutions for $\lambda = 2 \pm \frac{2}{9}\sqrt{6}$ and it has
three solutions for $2 - \frac{2}{9}\sqrt{6} < \lambda < 2 + \frac{2}{9}\sqrt{6}$ A graph of the set of
solutions as a function of λ looks roughly as in figure 2.

Bifurcation points are the points where the nature of the total set of solutions changes. Therefore there are two bifurcation points namely, the points $\lambda = 2 - \frac{2}{9}\sqrt{6}$ and $\lambda = 2 + \frac{2}{9}\sqrt{6}$

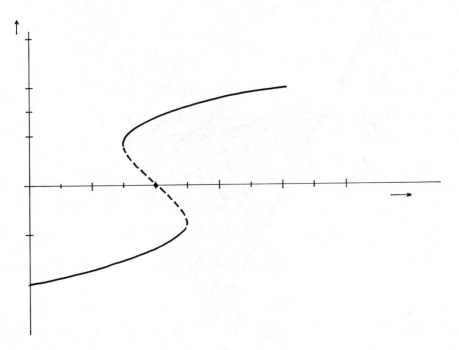

Figure 2.

Below is a picture (Figure 3) of $E(\lambda,x)$ as a function of both λ and x in which one can see how $E(\lambda,x)$ changes as λ varies.

Viewed from a different angle $E(\lambda,x)$ looks as below in Figure 4.

Here the dotted line indicates the equilibrium points: that is the bottoms of valleys and the tops of hills (see also figure 2!). As we walk through this two dimensional landscape in the direction of increasing λ we are first in a simple valley; then a new valley starts somewhere on the right hand slope so that shortly after we see on our right a moderate hill followed by the new valley followed by the original — so to speak — slope; going further still, the original valley peters out and only the new valley survives so that in the end we are again in a simple valley.

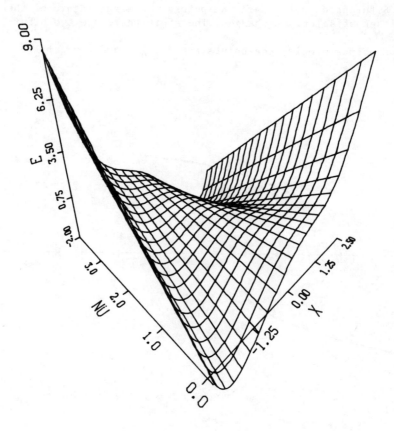

Figure 3.

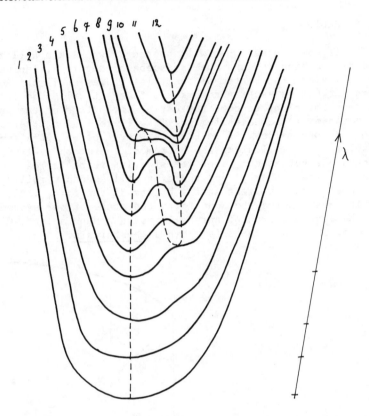

Figure 4.

3. STABILITY

Consider a ball in the landscape described by $E(\lambda,x)$. Its
dynamics will be given by $\ddot{x} = -E_x(\lambda,x) = -(x^3 - 2x - (2\lambda-4))$.
An equilibrium point x_0 will be called stable if for x' near
enough x_0 the solution of this differential equation, starting
in x' at time t=0 will remain near x_0 for all $t > 0$.
     Obviously in our example the equilibrium points given by
the solid lines in figure 2 are stable (the bottoms of the valleys) and the ones given by the dotted part are unstable (tops
of hills).

4. HYSTERESIS

It is easy to see what will happen if we kick a ball along
as we are walking through the landscape of figure 4. The ball
will roll along the bottom of the first valley until this one
peters out (cross section number 9), there it will suddenly
roll down the slope to come to rest at the bottom of the new
valley and as λ increases further it will roll along the bottom
of this second valley. If we then return, i.e. we let λ decrease, it will roll along the bottom of the second valley until this one peters out; i.e. its starting point is reached
(cross section number 4) and there it will suddenly roll down
the slope to come to rest at the bottom of the first valley.
Thus as λ moves back and forth the state of our system,
i.e. the rest position of our ball, moves around a loop, Cf.
figure 5. This phenomenon is known as hysteresis.

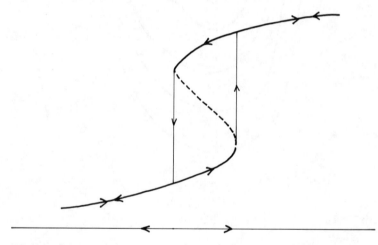

Figure 5.

5. EXAMPLE: A RUBBER BAR

Bifurcation and hysteresis phenomena like in the preceding
three sections occur in many guises. As another example consider a stiffish rubber bar with both ends fixed in such a position that it bulges slightly upwards (or downwards). Now let a
force F act on it in the transverse direction. The bar is supposed to be constrained to move in a plane. Positive F means it
tries to push the bar downwards: negative F tries to push it
upwards. As everybody knows (if not, it is a simple experiment
to do) a small force F will push the bar very slightly
downwards, as the force increases nothing much will

Figure 6.

happen until suddenly the bar snaps into the down position indicated by the dotted lines in Figure 6; diminishing the force
after the snap will not do much, indeed an equal negative force
is needed to make the bar snap back.

The potential energy of the bar at a given position x (of
its central point) with force F acting on it is definitely not
given by formula 2.1 with $F = \lambda-2$. But as far as a qualitative
description of its bifurcation, equilibrium points, stability
and hysteresis properties go, formula 2.1 gives a correct description.

6. AN ECONOMIC EXAMPLE: EQUILIBRIUM PRICES

This example is a simple modification of one described by
Wan in [7]. Consider an economy with two (groups of) traders 1
and 2 and two goods. The initial endowment of a trader of type
1 is $(t_1,0)$; the initial endowment of a trader of type 2 is
$(0,t_2)$.
Let the utility function of a trader of type 1 be

$U_1(x_1,x_2) = \min(3x_1^3 - 6x_1^2 + 4x_1, x_2)$ and the utility function

of a trader 2 be $U_2(x_1, x_2) = \min(3x_1,x_2)$.

There are as many traders of type 1 as of type 2 so that
the whole situation can be conveniently depicted in a two person, two good box diagram. Cf. figure 7 below. The point P =
(x,y) in the box represents the state of the economy where traders of type 1 have the bundle of goods (x,y) and the traders
of type 2 (t_1-x, t_2-y).
The Engel curve for traders of type 1 is the curved heavy line
starting in the lefthand corner (the graph of x_2

$= 3x_1^3 - 6x_1^2 + 4x_1$ with a horizontal segment in the upper

right hand corner added); the Engel curve for trades of type 2
is the slanted heavy dotted line starting in the upper right
hand corner. Given prices (p_1,p_2) consider the line making an

angle with the negative X-axis with $\mathrm{tg}\gamma = p_1 p_2^{-1}$. Then the

bundles of goods which are within the budget of trader 1 are to

the left of this line and the bundles within the budget of
trader 2 to the right of this line. As both Engel curves have
derivatives $\geqslant 0$ (viewed from the respective origins) the optimal
bundles of goods for traders 1 and 2 in the situation drawn in
figure 7 are respectively B_1 and B_2. Assuming that excess
demand for a good causes its price[2]

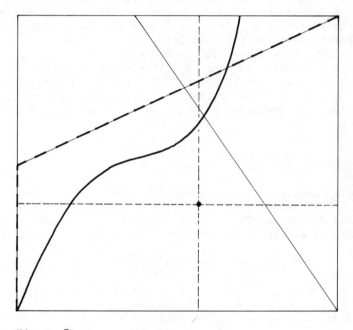

Figure 7.

to rise and vice versa, it is clear that the equilibrium states
of the economy are the points where the Engel curves intersect
(i.e. the point Q in the picture) and that the ratio of the
equilibrium prices is given by the angle of the line from this
point to R. Also such an equilibrium will be stable in the case
of an intersection point as in Figure 8a and unstable in the
case of Figure 8b.

Figure 8a   Figure 8b

Now consider how the equilibrium prices evolve as the economy grows and shrinks according to $t_1 = s$, $t_2 = \frac{2}{3}s$, $s \in [1,3]$.

Cf. Figure 9. Then for small s and big s there is precisely one stable equilibrium price and in between them is a region where there are three equilibrium prices of which one is unstable and two are stable. The graph of the equilibrium price ratios as a function of s is sketched in figure 10. As the economy grows and shrinks again we see hysteresis phenomena occur.

REMARK. In this case equilibrium states where one of the goods is free (i.e. has price zero) do not occur. In case both Engel curves are convex there can be no more than two intersection points in the interior of the box and free goods (at equilibrium) do tend to occur, Cf. Wan loc. cit. In this case the equilibrium prices bifurcation diagram looks something like in Figure 11 below.

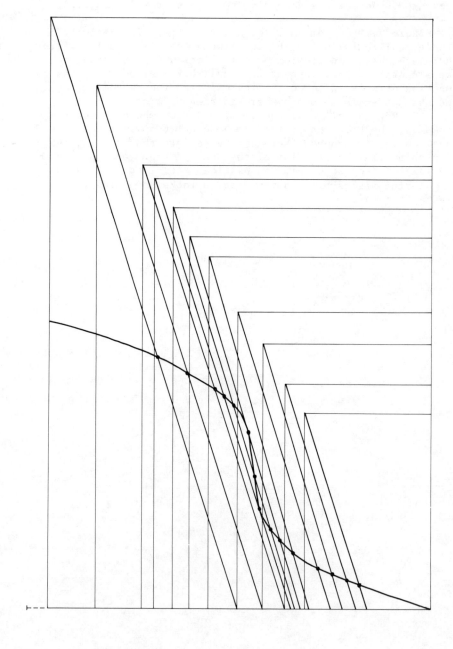

Figure 9.

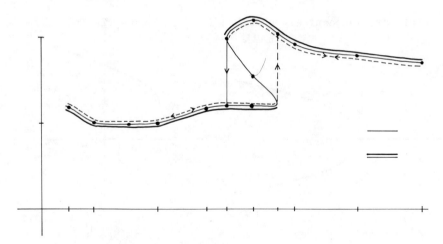

Figure 10.

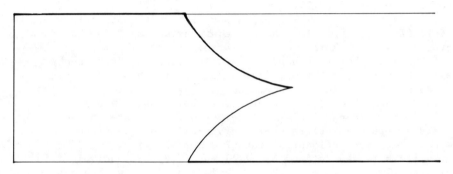

Figure 11.

7. THE PITCHFORK BIFURCATION

Now consider another family of potentials, namely

$$E(\lambda,x) = \tfrac{1}{4}x^4 - \tfrac{1}{2}\lambda x^2 \tag{7.1}$$

or slightly more generally

$$E(\lambda,x) = \tfrac{1}{4}x^4 - x^3 - \tfrac{1}{2}\lambda x^2 \tag{7.2}$$

The bifurcation diagrams for particles moving in these potentials are respectively sketched in the Figures 12a and b.

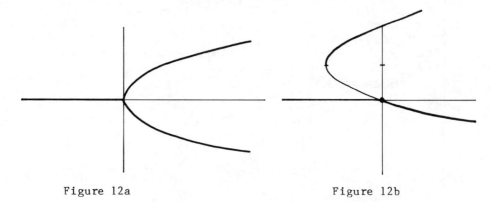

Figure 12a   Figure 12b

Given these bifurcation diagrams the reader will have no difficulty in sketching the corresponding landscapes as in Figure 4
for the potential (2.1).

Note that here as λ grows a particle in the potential
field (7.1) (or (7.2)) in stable equilibrium for $\lambda < 0$ has as λ
crosses zero a "choice" of equilibria: it can either follow the
left hand or the right hand valley. The smallest disturbance
will be decisive here. Or, from a more positive point of view,
if we are in a situation where we have extra control variables
(besides possibly λ) the values of λ where there is a pitchfork
bifurcation are particularly interesting. These are the hot
spots or pressure points where very small controls (invest-
ments, ...) are influential out of all proportion.

The surfaces defined by (7.1) and (7.2) are equivalent in
practically any sense of the word. Indeed (7.2) is obtained
from (7.1) by the co-ordinate transformation $x \to x$, $E \to E$, $\lambda \to
\lambda + 2x$. However from the bifurcation point of view the two po-
tentials are quite different. Then we are interested in the solution set of $E_x(\lambda,x) = 0$ as a function of λ.

8. PHYSICAL EXAMPLE OF A PITCHFORK BIFURCATION

Consider again a stiffish rubber bar constrained to move
in a plane. Now let the right-hand-end be fixed and the left-
hand-end free to slide horizontally but otherwise fixed and let
a force $F = \lambda$ act horizontally on the left end. Cf. Figure 13.

For negative F the bar will remain straight. The positive
F it will either assume an upwards or downwards curved position. Although (7.1) does not give the potential energy of the
bar in position x (the height of the mid-point of the bar above

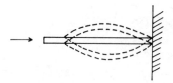

Figure 13.

the rest position) with force $\lambda = F$ acting on it, it does give
(more or less) the right qualitative description of the phenomena.

Actually a small positive force F will simply shorten the
bar a very slight bit and there will be a threshold which has
to be passed before the bar suddenly snaps into either of the
curved positions. So the true bifurcation diagram is more like
the one in figure 14 below.

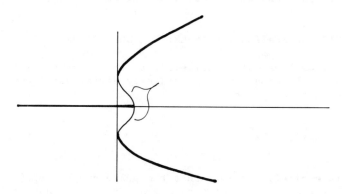

Figure 14.

9. GENERIC BIFURCATIONS

There are very few landscapes on earth which have the
hill-valley configuration corresponding to the bifurcation diagrams of Figure 12. The reason is simple. These are not structurally stable patterns. This means that a small change in the
potential $E(\lambda,x)$ will give a qualitatively different bifurcation diagram. E.g. if ε is small the perturbation

$$E(\lambda,x) = \tfrac{1}{4}x^4 - x^3 - \tfrac{1}{2}\lambda x^2 + \varepsilon x \tag{9.1}$$

yields the bifurcation diagrams depicted in figure 15 below
(the dotted lines give the diagram for $\varepsilon=0$; cf. Figure 12b).

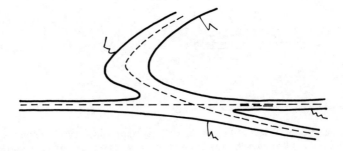

Figure 15.

This makes it important not only to study possible bifurcation but also deformations of bifurcations, that is families
of bifurcating systems of equations. Some of the buzz words
here are (universal) unfoldings of singularities, catastrophe
theory (cf. [8]) and imperfect bifurcations.

10. DYNAMIC BIFURCATIONS. EXAMPLE: THE HOPF BIFURCATION

So far we have only considered static bifurcations so to
speak. Now consider a family of ordinary differential equations
(or difference equations)

$$\dot{x} = f(x, \lambda), \quad x \in R^n \tag{10.1}$$

Then it is entirely possible that the phase diagram of (10.1),
i.e. the diagram of all solution curves of (10.1) changes in
nature as λ varies. It is e.g. possible that for $\lambda \leqslant 0$ the
phase diagram looks like Figure 16a (there is an equilibrium
point which is a stable attractor) while for $\lambda > 0$ the equilibrium point becomes unstable and there is a stable limit cycle

given by $x^2 + y^2 = \lambda$. Cf. Figure 16b. This is a so called Hopf-
bifurcation where a periodic solution bifurcates from a stationary one. A set of equations exhibiting this Hopf bifurcation
is

$$\dot{x} = -y - x (x^2 + y^2 - \lambda)$$
$$\dot{y} = x - y (x^2 + y^2 - \lambda) \tag{10.2}$$

Hopf bifurcations occur very often (cf. e.g. [5]). A very simple example where it occurs is for an electrical RLC loop with
a non-linear resistor, cf. [3, section 10.4].

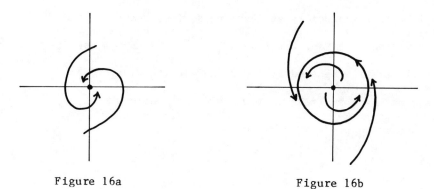

Figure 16a   Figure 16b

11. RECOGNISING BIFURCATION POINT CANDIDATES

Consider a family of equations

$$F(x,\lambda) = 0 \tag{11.1}$$

where for simplicity $x \in R^n$, $\lambda \in R$. Here $F: R^{n+1} \to R^n$ is supposed to be differentiable. Assume that $(0,\lambda_0)$ is a solution.
Consider $F_x(0,\lambda_0)$. If $F_x(0,\lambda_0) \neq 0$, then the implicit function
theorem says that locally we can solve for x in terms of λ so
that locally around $(0,\lambda_0)$ the solutions are a single valued
function of λ. Thus for a point like P in Figure 12a or R or Q
in Figure 12b to occur we need $F_x = 0$ (In the case of Figure
12a we have $F(\lambda,x) = x^3 - \lambda x$, $F_x(\lambda,x) = 3x^2 - \lambda$ so that indeed
$F_x(0,0) = 0$; in case of Figure 12b $F(\lambda,x) = x^3 - 3x^2 - \lambda x$,
$F_x(\lambda,x) = 3x^2 - 6x - \lambda$ so that $F_x(0,0) = 0$ and $F_x(2\frac{1}{4},1\frac{1}{2}) = 0$).

12. PITCHFORK BIFURCATION POINT THEOREM (VERY SIMPLE CASE). BI-
FURCATION FROM A SIMPLE EIGENVALUE

Consider again (11.1). Assume that $F(x,\lambda) = 0$, $F(0,\lambda) = 0$

for all λ. That is locally at least no serious restriction after a reparametrisation. Let $F_x(0,\lambda_0) : R^n \to R^n$ have a one dimensional kernel N spanned by $u \in R^n$ and let the range of
$F_x(0,\lambda_0)$ be given by $\{v \varepsilon R^n : \langle v,w \rangle = 0\}$ for some $w \in R^n$, $w \neq$
0. Now assume that

$$\langle u, F_{x\lambda}(0,\lambda_0) \ w \rangle \neq 0 \tag{12.1}$$

Then locally near $(0,\lambda_0)$ there is besides the branch $(0,\lambda_0 + t)$
of solutions a second branch through $(0,\lambda_0)$ of the form $(su +$
$s^2\alpha_1(s), \lambda_0 + s\alpha_2(s))$, s small, for suitable continous functions α_1 and α_2 of s.

In case n = 1 condition (12.1) simply becomes $F_{x\lambda} (\lambda,x) \neq$
0. Thus e.g. in the case of Figure 12a we have $F_{x\lambda} (0,0) = -1$
so that P = (0,0) is indeed a pitchfork bifurcation point; and
in case of Figure 12b $F_{x\lambda} (\lambda,x) = -1$ so that R is also a pitchfork bifurcation point. (Note that near Q the hypotheses of the
theorem are not fulfilled, also after translating x so that Q
becomes of the form $(0,\lambda_0)$).

13. HOPF BIFURCATION THEOREM

Let $F : R^n \times R \to R$ be analytic and let $F(0,\lambda) = 0$. Let
there be an eigenvalue $\sigma(\lambda)$ of $F_x(0,\lambda) : R^n \to R^n$, depending
smoothly on λ which crosses the imaginary axis at $\lambda = \lambda_0$ with
non-vanishing velocity, i.e. such that $R\sigma'(\lambda) \neq 0$, $\sigma(\lambda) = i\omega_0$,

$\omega_0 \in R$, $\omega_0 > 0$. Assume that the eigenvalue $i\omega_0$ is simple and that there are no eigenvalues of $F_x(0,\lambda_0)$ of the form $ik\omega$, $k \in Z \setminus \{1,-1\}$. Then

$$\dot{x} = F(x,\lambda) \tag{13.1}$$

has a family of periodic solutions of the form $(u(\omega(\varepsilon)t, \varepsilon), \lambda(\varepsilon))$ with $u(s,\varepsilon)$ periodic of period 2π in s and $\lambda(0) = \lambda_0$, $\omega(0) = \omega_0$, $u(s,0) = 0$.

In the case of example (10.2)

$$F_x(0,\lambda) = \begin{pmatrix} \lambda & -1 \\ 1 & \lambda \end{pmatrix}$$

so that its eigenvalues are $\lambda \pm i$. And the Hopf theorem applies.

14. CONCLUDING REMARKS

The material described above only touches the outermost fringes of the subject. In particular there are far reaching generalisations of the two theorems in sections 12 and 13 above. Even in the simple cases described here there is more to be said. Also there exist non-local bifurcation theorems (the two given above are strictly local). A modern up-to-date high level textbook on bifurcation theory is [1]. For an introduction to the numerics of bifurcation the reader can e.g. consult [6]. A wealth of applications of bifurcation theory (and the related catastrophe theory cf. [8]) is contained in [2] and [4].

REFERENCES

1. Chow, S.N., Hale, J.K.: 1982, Methods of bifurcation theory, Springer.

2. Gurel, O., Rössler, O.E. (eds.): 1979, Bifurcation theory and applications in scientific disciplines, Ann. New York Acad. Sci., 316.

3. Hirsch, M.W., Smale, S.: 1974, Differential equations, dynamical systems and linear algebra, Acad. Pr.

4. Holmes, Ph.J. (ed.): 1980, New approaches to nonlinear problems in dynamics, SIAM.

5. Marsden, J.E., Mc Cracken, M.: 1976, The Hopf bifurcation and its applications, Springer.

6. Mittelmann, H.D, Weber, H. (eds.): 1980, Bifurcation problems and their numerical solution, Birkhäuser.

7. Wan, H.Y.: Causally indeterminate models via multi-valued differential equations. In [2], pp. 530-544.

8. Zeeman, E.C.: 1982, Bifurcation and catastrophe theory, in: R. Lidl (ed.), Papers in algebra, analysis and statistics, Amer. Math. Soc., pp. 207-272.

BIFURCATION AND CHOICE BEHAVIOUR IN COMPLEX SYSTEMS

André de Palma

C.O.R.E., Université Catholique de Louvain and
Servico de Chimie-Physique II,
Université Libre de Bruxelles

1. INTRODUCTION

The evolution of systems where brisk transitions may occur
has, in the last few years, attracted the attention of the eco-
nomist, the sociologist and even the business executive. These
discontinuous changes are often seen as choice behaviours at
the individual or collective level. In this paper we will deal
with the cases where the system's choices result from the aggregation of individual behaviours and can be interpreted in
terms of bifurcations (Nicolis and Prigogine, 1977, and Prigogine, 1979). However, we do not deny that this concept of bifurcation can also be applied at the individual level. The individuals we shall consider here are supposed to have interdependent behaviours so that the system's behaviour is not
merely reduced to the sum of individual behaviours, and collective behaviours are effectively observed. We note that in the
contrary cases the system is linear; as pointed out by Conlisk
(1976), linear systems have attracted model makers in Economics
and Social Sciences because of the simplicity of the analytical
solutions they provide as well as the richness of their statistical properties. Our starting point is microscopic in the
sense that we describe the various transitions from state to
state carried out by the individuals (entities) of the system
in the course of time. A transition can correspond to the acquisition of information, the contraction of a disease, and the
adoption of a behaviour or an idea (or forgetting of this
idea). Kinetic models describe such types of reactions or
transitions from one state to another[1]. Let us remark lastly
that the social systems introduced here have not taken account

M. Hazewinkel et al. (eds.), Bifurcation Analysis, 31–48.

so far of social structures (social networks, inter- and intra-
group dependency, kinship, etc.) which undoubtedly play an important role in a large number of social transformation processes, as well as in so many other fields! We now describe the
framework of the models discussed.

Let us consider a system made up of N individuals who can
each occupy 1 out of J states. Two kinds of behaviour are taken
into account: the exchange of information and the transitions
from one state to another. The probability that an individual
moves from state i toward state j ($\neq$i; i,j=1,2,...N) during the
time interval (t,t+Δt), depends on the whole on the history of
the system (that is, on the various transitions carried out and
the information received until instant t), so that we have to
take account of memory effects. The corresponding evolution
equations are therefore non-Markovian and consequently very
difficult to manipulate. We shall propose two series of hypotheses giving rise to analytically tractable models and corresponding to the two limit cases described here. We shall assume
that in both cases the individuals (of the same community) are
indistinguishable, that is, they have statistically identical
behaviours.

In the first limit case, it is assumed that the individuals transit between choices only when they meet each other. The
meeting rates between two (or more) individuals depend on the
global state of the system and the transitions from one state
to another for these individuals depend only on their present
state. Moreover, the meeting durations are small when compared
to the time between two meetings in the system, so that the
global state of the system is clearly defind at any instant.
The model is therefore Markovian and the principal problem is
concerned with the construction of the probability of contact
between individuals. This limit case has been widely treated
in Chemical Physics (Nicolis and Prigogine, 1977). In Social
Sciences such a kinetic model has been used to describe for
example the diffusion of an innovation (Batholomew, 1973, and
Karmeshu and Pathria, 1980) or of a new product (Bass, 1969):
in the simplest case, there are two possible states for each
individual, "aware" or "not aware" of the innovation. Another
application is the study of typically collective phenomena such
as rioting (Burbeck et al., 1978). In a somewhat different context, the propagation of an epidemic can be described by a model of this type (see Bailey, 1975, for a review of recent
works in this field); for the classical simple epidemic model
e.g. the two possible states for each individual are "susceptible" and "infected". It is worth noting that although
Bernoulli (1760) has already expressed in mathematical terms a
problem in epidemiology, it was only much later that Hammer
(1906) used a kinetic approach for a "chemical" type. In most
works, special attention is paid to the discussion of the conditions for the growth of a given variable and the existence of

a threshold. Strangely enough, in Marketing very few authors (Glaister, 1974, and Bemmaor, 1980) discussed the presence of bifurcation and the condition of existence of a threshold.

In the second limit case, the processes of transmission of information are supposed to occur sufficiently often in comparison with the processes of transition from state to state so that each individual knows the state of the system. On the basis of this information individuals can decide to review, at any instant, their actual state, and in case of need transit to a new state. The model thus becomes Markovian. This limit case has been considered by several authors (Coleman, 1964, Bartholomew, 1973, and Conlisk, 1976). It has been analysed in the frame of choice problems in a recent paper (de Palma and Lefèvre, 1981 and 1983).

In sections 2 and 3, we show the interest of these two limit cases by means of several models introduced in the literature. These will allow us to underline the relevance of the concept of bifurcation in Social Sciences. We note that the importance of this concept in this field has been illustrated previously by several authors, in particular in Regional Science by Isard and Liossatos (1977), Weidlich and Haag (1980), and Wilson (1981); and in Economics and Management Science by Coppock (1977), Smale (1979), Wan (1979) and de Greene (1982), for example. Some applications are presented briefly in section 4.

2. FIRST LIMIT CASE MODELS

The starting point of non-linear kinetic models is often the discussion of a differential equation similar to the following

$$dX/dt = a + bX + cX^2 + dX^3. \qquad (2.1)$$

Most work in Social Sciences which uses this equation does not justify its introduction from a microscopic point of view, only reference is made to Chemical Physics; its status in this case is totally phenomenological and its interpretation seems to us rather ambiguous.

Recently, by means of heuristic arguments, de Palma (1981) has derived equations of the type described in (2.1). It seems to us very interesting to present briefly the arguments involved; this approach is situated in a perspective close to that of the Brussels Group (Nicolis and Progogine, 1977). Consider a population of N individuals and suppose that each of them can belong to one of the two classes C_1 and C_2; a class corresponds, for example, to an opinion, a type of behaviour, etc. We can reasonably assume that each individual has a sphere of influence (Moles and Rohmer, 1977) and that an individual leaves a class

only if he belongs to the sphere of influence of an individual
from the opposite class[2]. Let X denote the number of individuals in class C_1; then the evolution of the number of "collisions" over time gives rise to the following differential
equation (de Palma, 1981)

$$dX/dt = f(d)N^{d-1}X(N-X), \quad \text{with } f(1) = \alpha, \quad (2.2)$$

in which d is a parameter linked to the sphere of action of
individuals; f(d) is a function of the parameter d, and measures the level of meeting and conversion of individuals. d=0
corresponds to the case of Chemical Kinetics: equation (2.2)
can be expressed solely in terms of density, the interactions
being thus local; when d=1, equation (2.2) can be expressed
solely in terms of absolute variables:

$$dX/dt = \alpha X(N-X), \quad (2.3)$$

which corresponds to the case where the sphere of influence of
individuals is large compared to the size of the system.

More realistic models need the introduction of supplementary interactions between individuals; so, for equation (2.3),
including higher order non-linear communication processes (Webber, 1972):

$$2I_X + I_Y \xrightarrow{\beta X^2} 3I_X, \quad (2.4)$$

and a forgetting process which corresponds to the introduction
of a "negative loop" (Bartholomew, 1973):

$$I_X \xrightarrow{\rho} I_Y \quad (2.5)$$

as well as a linear process of acquisition of opinion or behaviour (Bemmaor, 1980):

$$I_Y \xrightarrow{\theta} I_X \quad (2.6)$$

yields

$$dX/dt = (\alpha X + \beta X^2)(N-X) - \rho X + \Theta(N-X). \quad (2.7)$$

We have already studied this equation in detail in a previous
work (de Palma, 1981). We limit ourselves here to the presentation, in Figure 1, of certain results illustrating bifurcation
phenomena; their interpretation, rather intuitive, is left to
the reader.

When writing (2.4), (2.5) and (2.6), we based ourselves on the formalism used in Chemical Kinetics to describe molecular reactions. In reality, these processes are stochastic and therefore are better described by means of Markovian master equations than by differential equations of type (2.7). These master equations are on the whole very difficult to study analytically (Malek Mansour, 1979), hence approximation methods have been developed to study them. In particular Kurtz (1978) has proved various theorems which under certain conditions allow to associate with the stochastic version a deterministic version in terms of differential equations. The main condition required by Kurtz's theorem is the extensiveness of the transition probabilities. In chemistry, the interactions (molecular interactions) are local so that this condition is practically always fulfilled; for social systems this is not necessarily true. As an illustration, we now discuss an example due to Malek Mansour (1979) which shows that for long range interactions (i.e. d=1; this corresponds, for example, to radio or videophone communications as opposed to local communications of the type mouth-to-ear), the probabilistic and associate deterministic models may have opposed behaviours.

Let us consider the system introduced above with X individuals in class C_1 and N-X individuals in class C_2; they are supposed to move either spontaneously:

$$I_X \rightleftharpoons I_Y, \tag{2.8}$$

or in an induced way

$$I_X + I_Y \longrightarrow 2I_X, \tag{2.9}$$

$$I_X + I_Y \longrightarrow 2I_Y. \tag{2.10}$$

The parameter b is independent of N so that the interactions between individuals are long range interactions (non-local interactions). The deterministic differential equation corresponding to the process (2.8), (2.9) and (2.10) is

$$dX/dt = 2a(N/2-X). \tag{2.11}$$

For symmetry reasons, evolution of X is independent of parameter b and the stationary solution (defined by dX/dt=0) is equal to N/2. Malek Mansour (1979) has constructed and studied the Markov chain associated with the processes (2.8), (2.9) and (2.10). The main result is the following:
if b < a, i.e. if the non-linear interactions are weak, the stationary distribution is maximum in X=N/2, which corresponds to the stationary solution of the macroscopic equation (2.11); if, on the contrary, b > a, the stationary distribution is maximum in X=0 or X=N, and thus the behaviour of the system is

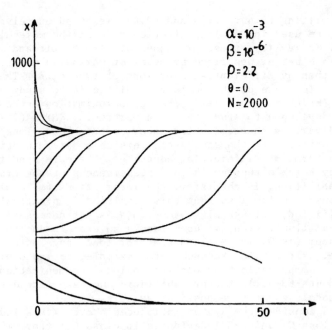

Figure 1(a). Temporal evolutions

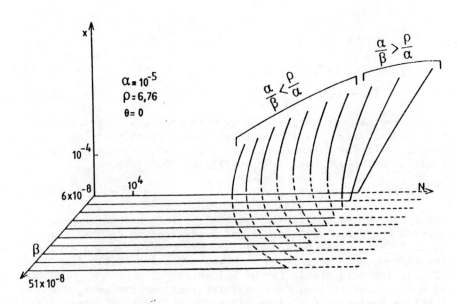

Figure 1(b). Bifurcation diagram; N and β bifurcation parameters

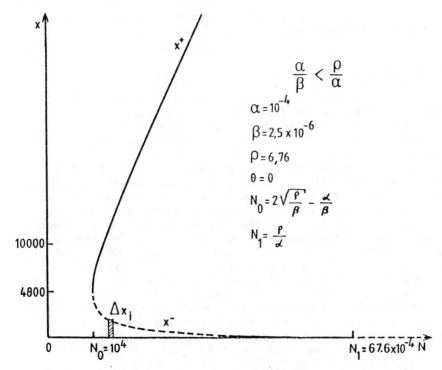

$$\frac{\alpha}{\beta} < \frac{\rho}{\alpha}$$

$\alpha = 10^{-4}$

$\beta = 2{,}5 \times 10^{-6}$

$\rho = 6{,}76$

$\theta = 0$

$N_0 = 2\sqrt{\frac{\rho}{\beta}} - \frac{\alpha}{\beta}$

$N_1 = \frac{\rho}{\alpha}$

Figure 1(c). Bifurcation diagram; N bifurcation parameter

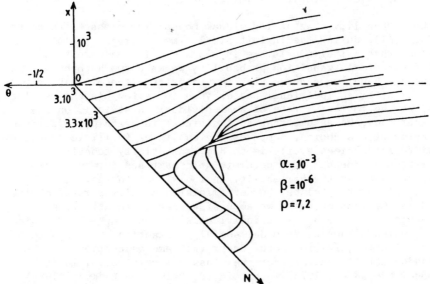

$\alpha = 10^{-3}$

$\beta = 10^{-6}$

$\rho = 7{,}2$

Figure 1(d). Bifurcation diagram; N and Θ bifurcation parameters

qualitatively new with regard to the results given by the
macroscopic description.

3. SECOND LIMIT CASE MODELS

In this section we would like to introduce the study of
the choice behaviour of interacting individuals. This kind of
model has been studied in detail in a previous article (de
Palma and Lefèvre, 1983), and we shall give only a brief
account of results which exhibit bifurcating behaviours. The
models presented borrow from Psychology (Luce, 1959) and Economics (McFadden, 1973) the concept of utility and the hypothesis
that the individual choices are derived from the maximisation
of a random utility function. Moreover, they are connected with
Sociology in the sense that they try to express the interdepen-
dency between individual decisions and the dynamic elements in
decision making (Conlisk, 1976, and Schelling, 1978). In the
introduction we have already underlined the importance of
interdependence between the units of a system.

Let us consider the system introduced in section 2 of N
individuals who can be either in class C_1 or in class C_2; each
class will correspond, here, to a particular choice. Let $X_i(t)$,
i=1,2 be the number of individuals in class C_i at time t; we
will compute the time evolution of $P_{x_o,x}(t)$, the probability
that $X_1(t)=x$ given that $X_1(0)=x_o$. Let us define by

$$R_{ij}(x)\Delta t + o(\Delta t), \qquad i=1,2, \quad j\neq i, \qquad (3.1)$$

the probability that an individual moves from class C_i to class
C_j, $j(\neq i)$, during the infinitesimal time interval $(t,t+\Delta t)$,
given that x individuals are in class C_1. To evaluate this
infinitesimal transition probability we make two hypotheses. On
the one hand, we suppose that the transition $i\rightarrow j(\neq i)$ may only
occur after an individual has decided to review his choice; let
R_i be the reviewing rate of an individual in state i. On the
other hand, we assume that an individual in state C_i, after the
review of his choice, compares the qualities (utilities) of the
different choices available to him; the utility function is
written as the sum of a deterministic part and a random term
which measures the variability in individual tastes. Let
$V_{j/i}(x)$ and $V_{i/i}(x)$ be the deterministic utility part of choice
C_j and C_i respectively; we emphasise that these utilities
depend on the number x of individuals who have opted for choice
C_1, which allows us to model social interaction phenomena such
as imitation, fashion, etc. Under the same assumptions on the
random utility part as for the classical multinomial logit
model (McFadden, 1973) we obtain (de Palma and Lefèvre, 1983)

$$R_{ij}(x) = R_i \frac{\exp[V_{j|i}(x)/\mu]}{\sum\limits_{s=1}^{\ell} \exp[V_{s|i}(x)/\mu]} , \qquad \begin{matrix} i=1,2 \\ j \neq i \end{matrix} \qquad (3.2)$$

where μ is a parameter related to the importance of the random utility part and thus measures the fuzziness in the individual decision making (this is illustrated in Figure 2). The time evolution of $P_{x_o,x}(t)$ is given by the following prospective Kolmogorov equation (Parzen, 1962)

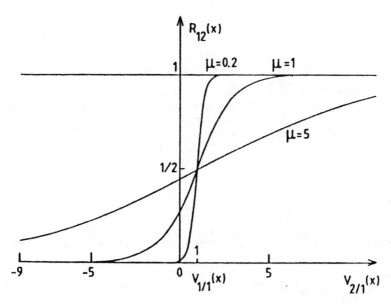

Figure 2. Transition probability $R_{ij}(x)$ from class C_i to class C_j

$$\frac{dP_{x_o,x}(t)}{dt} = - R_1 \left\{ \frac{x \exp[V_{2\ 1}(x)/\mu]}{\exp[V_{1|1}(x)/\mu] + \exp[V_{2|1}(x)/\mu]} \right.$$

$$+ R_2 \left. \frac{(N-x)\exp[V_{1\ 2}(x)/\mu]}{\exp[V_{1|2}(x)/\mu] + \exp[V_{2|2}(x)/\mu]} \right\} P_{x_o,x}(t)$$

$$+ R_1 \frac{(x+1)\exp[V_{2|1}(x+1)/\mu]}{\exp[V_{1|1}(x+1)/\mu] + \exp[V_{2|1}(x+1)/\mu]} P_{x_o,x+1}(t)(1-\delta_{x,N})$$

$$+ R_2 \frac{(N-x+1)\exp[V_{1|2}(x-1)/\mu]}{\exp[V_{1|2}(x-1)/\mu] + \exp[V_{2|2}(x-1)/\mu]} P_{x_o,x-1}(t)(1-\delta_{x,o})$$

$$(3.3)$$

Equation (3.3) has been studied for different utility functions $V_{i|i}(x)$ (de Palma and Lefèvre, 1983); we will describe here very briefly the case where the deterministic utility part of a choice is a linear function of the number of individuals who have opted for this choice, i.e.

$$V_{1|1}(x) = V_{1|2}(x) = a + bx, \qquad\qquad (3.4)$$
$$X = 0, \ldots, N$$
$$V_{2|1}(x) = V_{2|2}(x) = c + d(N-x)$$

The stationary probability distribution is easily found to be equal to

$$P_x = \lim_{t\to\infty} P_{x_o,x}(t) = \Pi_x / \sum_{s=o}^{N} \Pi_s, \qquad x = 0, \ldots, N \qquad (3.5)$$

where $x_o=1$, and for $x=1, \ldots, N$,

$$P_x = \binom{N}{x}(R_2/R_1)^x \exp\left[(a-c-dN)(x+1) + (b+d)x(x+1)/2]/\mu\right\} \cdot$$

$$\frac{[1+\exp[c+dN-a-(b+d)x]/\mu}]}{[1+\exp(a-c-dN)/\mu}]} \qquad (3.6)$$

As an illustration, let us examine the symmetrical case where $a=c$, $b=d$ and $R_1=R_2$. In this case, the stationary probability distribution is symmetric and can be either monomodal or bimodal; it can be shown that the condition for having a bimodal distribution is

$$bN \geqslant (\mu/2)\ln[1+4/N]^N . \tag{3.7}$$

This is illustrated in Figure 3 for large N. Thus, for N given, when the non-linear effects measured by parameter b are strong enough, the individuals can aggregate around well differenti-ated preferences; in other words, when parameter b is large enough, the system "chooses" one of the two states C_1 or C_2 (these, for $x_0=N/2$, have equal chances to be selected). A similar result has been shown in the model of Malek Mansour discussed in section 2. We note that b can be interpreted as a coefficient of imitation (b>0) or anti-imitation (b<0). Now for b given, an interpretation in function N is also interesting. We see that the stationary choice distribution is qualitatively modified in function of the value of N: for N large enough, the state X=N/2 becomes the less probable whilst it is the most probable for small values of N.

Oppenheim, Schuler and Weiss (1977) have studied, with Markovian master equations, the time evolution of two particu-lar chemical systems. In their study they have introduced a distinction between quasi-stationary states reached after a time $O(N)$ where N measures the size of the reaction system), and the true equilibrium state reached after a supplementary

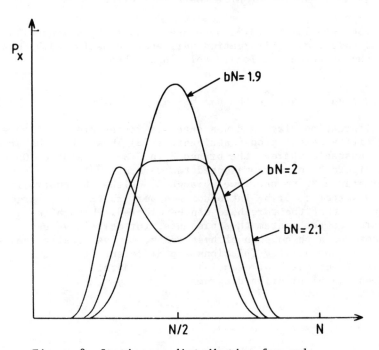

Figure 3. Stationary distribution for $\mu=1$.

time $O(e^N)$. The stationary state corresponds to a bimodal dis-

tribution and the quasi stationary state to monomodal distribu-
tions centred at these two modes; we hope to be able to show
similar results for the model described by equation (3.3) and
(3.4) and thus give a more precise meaning to the expression
"choice of the system".

In the article referred to above (de Palma and Lefèvre,
1983), we have followed Kurtz (1978) and introduced the deterministic version associated with the stochastic model (3.3).
Let $Z(t)$ be the proportion of individuals in state C_1 at time
t; let us assume that the utilities are linear function of Z,
i.e.

$$V_{1|1}(Z) = V_{1|2}(Z) = a + bZ, \qquad\qquad (3.8)$$
$$0 < Z < 1$$
$$V_{2|1}(Z) = V_{2|2}(Z) = c + d(1-Z) .$$

We showed that the deterministic version of model (3.3), when
$R_1 = R_2 = R$, is

$$\frac{dZ}{dt} = R \left(\frac{\exp[(a+bZ)/\mu]}{\exp[(a+bZ)/\mu] + \exp[(c+d(1-Z))/\mu]} - Z \right) . \qquad (3.9)$$

Equations similar to (3.9) have been studied by Wilson (1981).
Let us take c as a bifurcation parameter; equation (3.9) has
then one or three stationary solutions. When

$$B > 4\mu \qquad\qquad \text{with} \quad B = b + d \qquad\qquad (3.10)$$

the bifurcation diagram has a shape as represented in Figure 4.
A stability calculation (Sattinger, 1973) shows that the branch
Z_0 is unstable whereas the branch Z_+ and Z_- are stable. On the
same figure we have represented two ways of moving from the
upper branch Z_+ to the lower branch Z_-: either by controlling
the parameter c, or by introducing a perturbation ΔZ_i large
enough so that the threshold can be cleared. Through the non-
linearities the system thus presents, for certain values of its
parameters, a phenomenon of hysteresis. Let us finally remark
that when the utility functions are different (logarithmic
utility functions, for example), one can obtain other
interesting bifurcation diagrams.

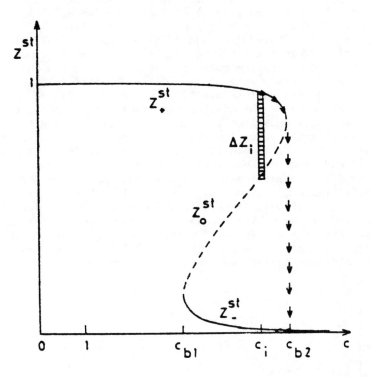

Figure 4. Bifurcation diagrams; c parameter of bifurcation

4. APPLICATIONS

In this paper we have presented some mathematical tools and modelling approaches which have been used by different members of the Brussels Group to study social systems.

So, in Biology and Ecology we would like to mention Deneubourg and Allen (1976), Goldbeter and Nicolis (1976) and Lefever, Nicolis and Prigogine (1967). In the same spirit as the work presented here, interesting studies have also been done to model the choice behaviour of some social insects (Deneubourg, Pasteels and Verhagen, 1981): these authors have analysed the behaviour of ants in a nest with regards to some choices such as the search of a source of food, for example.

Other work has been done in urban dynamics (Allen et al., 1981, Allen and Sanglier, 1981, and Ben-Akiva, and de Palma 1983). These studies essentially construct scenarios which

present interesting spatial structurations which illustrate the
concept of "order through fluctuations"; the existence of critical points (bifurcation points) in the evolution of such
systems will obviously interest the decision maker.

Similar models have also been developed in the field of
transportation (Herman and Prigogine, 1979). The modal choices
(choice of a transportation mode such as the car or the bus,
for example) illustrate some of the ideas presented in this
article: indeed, the options available to individuals are
interdependent because the utility functions corresponding to
each mode strongly depend on the number of users of this mode
(congestion effects, imitation phenomena, etc.). In particular
models of choice between bus and car (Deneubourg, de Palma and
Kahn, 1979 and Kahn Deneubourg and de Palma, 1981) have been
developed. Figure 5 represents a diagram in phase space for a
choice model recently introduced (Kahn et al., 1981): y represents the number of bus users and L the global level of services offered by the public transportation mode (bus). Two
stationary states are physically acceptable: $(0,0)$ corresponds
to the rejection of the public service and (y^+,L^+) corresponds
to the coexistence between the users of the bus and the users
of cars.

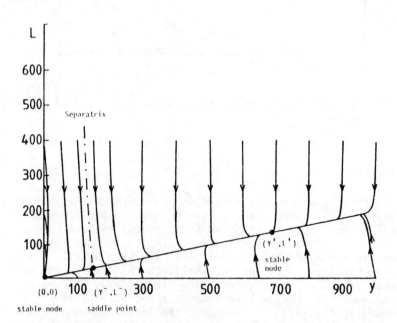

Figure 5. Trajectories in the (L,y) phase space, with one
stable node at $(0,0)$, one stable node at (y^+,L^+) and
one saddle point at (y^-,L^-); from Kahn et al. (1981)

Finally let us mention some theoretical work on situation in which the behaviour of the individuals results from expectations (de Palma, 1982).

We end this section by noting that the stochastic (and the deterministic) processes can be simulated by classical methods (Gillespie, 1977); empirical studies are envisaged in the near future to estimate the parameters of the models.

5. CONCLUSIONS

The application of bifurcation theory to social systems is at its very beginning. In this paper we have outlined briefly some attempts to model social systems which admit bifurcating behaviours. The orientation of this paper was methodological: we have tried to construct in a rational way kinetic and choice models and discussed some difficulties of this new apporach. Although empirical and theoretical problems remain, we are convinced of the interest of developing models which allow a better understanding and perhaps forecasting of the dynamic behaviour of complex systems.

NOTES

1) The "hydraulic" models in economics, presented by Fisher (a student of the famous physical chemist W. Gibbs) in his Ph.D. thesis, give an example of a linearised version of the kinetic models mentioned above; Fisher represented the various sectors of an economy by water tanks and studied the impact of water flows on the system.

2) This is a caricature description inspired mainly by the study of molecular collisions in chemistry.

REFERENCES

Allen, P.M. and Sanglier, M.: (1978), Dynamic Models of Urban Growth, Journal of Social Structure, 1, pp. 265-280.

Allen, P.M., Sanglier, M., Boon, F., Deneubourg, J.L. and de Palma, A.: (1981), Models of Urban Settlement and Structure as Self-Organising Systems, Final Report, Systems Analysis Division, D.O.T., Washington.

Bailey, N.T.J.: (1975), The Mathematical Theory of Epidemics, Charles Griffin, London.

Bass, F.M.: (1969), A New Product Growth Model for Consumer Durables, Management Science, 15, no. 5, pp. 215-227.

Batholomew, D.J.: (1973), Stochastic Models for Social
Processes, Wiley, London.

Bemmaor, A.C.: (1980), Threshold Effects of Advertising: a
Switching Regression Model, Internal report, Ecole
Supérieure des Sciences Economiques et Commerciales, Cergy
Pontoise.

Ben-Akiva, M. and de Palma, A.: (1983), Modelling and Analysis
of Dynamic Residential Location Choice, W.P. No. 83-19, Mc.
Master University.

Bernoulli, D.: (1760), Essai d'une nouvelle analyse de la
mortalité causée par la petite vérole, et des avantages de
l'inoculation pour la prévenir, Mémoire de Mathématique et
Physiquede l'Académie Royale de Sciences, in Histoire de
l'Académie Royal des Sciences, 1766, Paris.

Burbeck, S.L., Raine, W.J. and Abudu Stark, M.J.: (1978), The
Dynamics of Riot Growth: An Epidemiological Approach,
Journal of Mathematical Sociology, 6, pp. 1-22.

Coleman, J.S.: (1964), Introdution to Mathematical Sociology,
The Free Press, New York.

Conlisk, J.: (1976), Interactive Markov Chains, Journal of
Mathematical Sociology, 4, pp. 157-185.

Coppock, R.: (1977), Decision-Making when Public Opinion
Matters, Policy Science, 8, pp. 135-146.

Deneubourg, J.L. and Allen, P.M.: (1976), Modèles théoriques de
la division du travail dans les sociétés d'insectes,
Académie Royale Belgique, Bulletin de la Classe des
Sciences, 62, pp. 416-429.

Deneubourg, J.L., de Palma, A. and Kahn, D.: (1979) Dynamic
Models of Competition between Transportation Modes,
Environment and Planning, A, 11, pp. 665-673.

Deneubourg, J.L., Pasteels, J. and Verhaegen, J.Cl.: (1981),
Compte rendu de l'Union Internationale des Insectes Sociaux,
Toulouse.

de Palma, A.: (1981), Modèles Stochastiques des Comportements
Collectifs dans les Systèmes Complexes, Ph.D. thesis,
University of Brussels.

de Palma, A.: (1983), Incomplete Information, Expectation and Subsequent Decision Making, Journal of Environment and Planning, A, 15, pp. 123-130.

de Palma, A., and Lefèvre, Cl.: (1981), Simplification Procedures for a Probabilistic Choice Model, Journal of Mathematical Sociology, 6, pp. 43-60.

de Palma, A., and Lefèvre, Cl.: (1983), Individual Decision-Making in Dynamic Collective Systems", Journal of Mathematical Sociology, 9, pp. 103-124.

Gillespie, D.T.: (1976), A General Method for Numerically Simulating the Stochastic Time Evolution of Coupled Time Evolution of Chemical Reactions, Journal of Computation Physics, 22, pp. 403-434.

Glaister, S.: (1974), Advertising Policy and Returns to Scale, Economica, 41, pp. 139-156.

Goldbeter, A. and Nicolis, G.: (1976), Progress in Theoretical Biology, Vol. 4, Academic Press, New York.

Greene, K.B.: de (1982), The Adaptative Organization, Wiley, New York.

Hammer, W.N.: (1906), Epidemic Disease in England - The Evidence of Variability and the Persistence of Type, The Lancet, II, pp. 733-739.

Herman, R. and Prigogine, I.: (1979), A two-fluid approach to town traffic, Science, 204, pp. 148-151.

Isard, I. and Liossatos, P.: (1977), Models of Transition Processes, Papers Regional Science Association, 39, pp. 27-59.

Kahn, D., Deneubourg J.L. and de Palma, A.: (1981), Transportation Mode Choice, Environment and Planning A, 13, pp. 1163-1174.

Karmeshu, and Pathria, R.K.: (1980), Stochastic Evolution of a Non-linear Model of Diffusion of Information, Journal of Mathematical Sociology, 7, pp. 59-71

Kurtz, T.G.: (1978), Strong Approximation Theorems for Density Dependant Markov Chains, Stochastic Processes and their Applications, 6, pp. 223-240.

Luce, R.D.: (1959), Individual Choice Behaviour, Wiley, New
    York.

Malek Mansour, M.: (1979), Fluctuations et transitions de phase
    de non-équilibre dans les systèmes chimiques, Ph.D. Thesis,
    University of Brussels.

McFadden, D.: (1973), Frontiers in Econometrics, Academic
    Press, New York.

Moles, A. and Rohmer, E.: (1977), Théorie des Actes, Casterman,
    Paris-Tournai.

Nicolis, G. and Prigogine, I.: (1977), Self-Organization in
    Non-Equilibrium Systems, Wiley, New York.

Oppenheim, I., Schuler, K.E. and Weiss, G.M.: (1977),
    Stochastic Theory of Nonlinear Rate Processes with Multiple
    Stationay States, Physica 88A, pp. 191-214.

Parzen, E.: (1962), Stochastic Processes, Holden-Day, San
    Francisco.

Prigogine, I.: (1979), From Being to Becoming, Freeman, San
    Francisco.

Sattinger, D.: (1973), Topics in Stability and Bifurcation
    Theory, Springer Verlag, Berlin.

Smale, S.: (1979), On Comparative Statics and Bifurcation in
    Economic Equilibrium Theory, in: Bifurcation Theory and
    Applications in Scientific Disciplines, Annals of the New
    York Academy of Sciences, 316, New York.

Wan, H.Y.: (1979), Causally Indeterminate Model Via Multi-
    Valued Differential Equations, in: Bifurcation Theory and
    Applications in Scientific Disciplines, Annals of the New
    York Academy of Sciences, 316, New York.

Webber, M.J.: (1972), Impact of Uncertainty on Location, MIT
    Press, Cambridge, Mass.

Weidlich, W. and Haag, G.: (1980), Migration Behaviour of Mixed
    Population in a Town, Collective Phenomena, 3, pp. 89-112.

Wilson, A.G.: (1981), Catastrophe Theory and Bifurcation:
    Applications to Urban and Regional Systems, Croom Helm,
    London.

B. Applications

SOME REMARKS ON THE NATURE OF STRUCTURE AND METABOLISM IN
LIVING MATTER

B. Leijnse

Faculty of Medicine
Erasmus University Rotterdam

1. INTRODUCTION

The bifurcation approach to fundamental problems in (bio)-
chemistry and biology is relatively new. Although O. Gurel's
statement at the Eight FEBS MEETING in Amsterdam in 1972 that
bifurcation theory had not been implemented in the literature
should be corrected[1], he was quite right in his remark that
there did not exist a complete bifurcation analysis of dynamic
systems at the molecular level. Even now such a theory still
has to be developed; moreover it must be concluded that bifurcation, or peeling, or branching theory has up till now only
slightly affected (bio)chemistry and biology[2].
    It is therefore remarkable that the famous biochemist and
sinologue Joseph Needham added the following picture (figure 1)
to his well-known book Order and Life (1936, reprinted in
1967)[3]; it is clear from the text of the addendum that he
expected important results from the application of bifurcation
analysis to morphology, especially morphogenesis.
    But he was also aware of the severe criticism always ex-
pressed against this type of treatment of biological problems.
According to the critics biological models give more complicated solutions than the abstract models and still a biological
model is a crude simplification of the real situation in a
living organism. The physiologist J.S. Haldane (1932) wrote[4]:
"We are unable to apply mathematical reasoning to life since
mathematical treatment assumes a separability of events in
space which does not exist for life as such. We are dealing
with an invisible whole when we are dealing with a life". J.
Needham countered: "To close the door in this way to the possi-

51

M. Hazewinkel et al. (eds.), Bifurcation Analysis, 51–66.
© *1985 by D. Reidel Publishing Company.*

Figure 1. From: Order and Life by Joseph Needham with kind
           permission of Yale University Press
           View of a marshalling yard (London and North Eastern
           Railway) illustrating by analogy the concept of
           restriction of potentiality by multiple bifurcation.
           The photograph is taken from above the "hump". Up to
           this the freight wagons are pushed and from it there
           they run down individually over electric retarders to
           a number of alternative sidings, where they are ready
           to be despatched to a fresh destination. The top of
           the "hump" corresponds to the totipotent or maximally
           unstable condition of the egg.

bilities of the integral calculus, Topology and Mengenlehre
applied to biology is a gratuitous denial of the intellect".
But although this kind of counterattack is indispensable, it is
profitless as Haldane's stronghold is the axiom that scientific
biology depends on the existence of life as such.

2. VITALISM

As the complexity of living organisms is notorious and life is a very intriguing phenomenon, it is no wonder that the doctrine of vitalism is always active in the human mind. According to this doctrine, life in living organisms is caused and sustained by a vital principle that is distinct from all physical and chemical forces and cannot ultimately be explained in terms of physics and chemistry and that life is, in part, self-determining and self-evolving. An excellent example of vitalistic thinking is the conclusion drawn by R. Brown (1828)[5] from the uninterrupted and irregular "swarming" motion he microscopically detected with all kind of organic substances[5]; he believed he had found in these particles the "primitive molecule" of living matter. After more experiments, with inorganic material, he changed his opinion, but could not offer an explanation of the phenomenon. Since then it has been shown by eminent scientists, such as A. Einstein (1926)[6] that the Brownian movement is performed by bodies of miscroscopically visible size suspended in a liquid on account of the molecular motions of heat and has nothing to do with life; on the contrary, the study of Brownian motion contributed substantially to the development of physics and chemistry.

The next example of vitalism is much more important, and its influence is still present. Before chemistry was engaged in organic material, it was generally accepted that this material could never be synthesized in vitro, as a vital force was thought to be necessary. The urea synthesis by F. Wöhler (1828) and other synthesis of organic substances proved that a Vis Vitalis was superfluous. But nevertheless C. Bernard[7], who founded physiology and critisised vitalism severely, wrote: "It will be possible to produce by chemical synthesis the compounds of living organism, but it will never be possible to produce by chemical synthesis the instruments of living organisms because they are the products of the organised morphology which is outside the realm of chemistry and it will be no more possible for a chemist to produce an enzyme than to produce a living organism".

Here vitalism found a congenial abode in the form and was never totally expelled. The distinction between "forma" and "materia" is mainly due to Aristotle and had an enormous influence on the historical development of biology. Through the centuries the word form (forma) has continually undergone changes of meaning; sometimes it became abstract and possessed only a logical meaning, sometimes it meant external spatial shape. F. Bacon stated in 1600, "the form of a thing is its very essence", but later whatever was trivial was called a formality; however, in the study of living organism their form, their external spatial shape and contour, was and is of the utmost significance. The study of heredity and C. Darwin's evolutionary

theory were mainly based on the organic forms. Form is insepar-
able from individuality in man; and from the process of recognition it may be concluded that the form of an individual is to
a high degree an invariant. For many centuries past morphologists have studied the form of animals and plants without much
consideration of the matter. Form, however, is not the absolute
prerogative of the morphologist because no matter can exist
without form. Form is also an essential characteristic of the
whole realm of chemistry. It seems worthwhile to apply the
methods of physical science in order to clarify the problem of
organic form. Although chemistry and physics have already
solved many biological problems, a physical scientist nevertheless must always seriously consider the possibility that he
will suddenly find himself out of his depth when penetrating
deeper and deeper into the secrets of living matter. Experience
teaches that if a scientist unexpectedly loses his grip on a
biological problem the danger of vitalistic reasoning is imminent. Therefore it seems wise to outline the nature of the theories and methods applied in order to prevent sudden embarrass-
ments. Some words on chemistry follow.

3. THE BASIS OF CHEMISTRY

     The basic principles of modern chemistry were developed in
the seventeenth century. R. Boyle propagated the idea that chemistry and natural philosophy were one and he stressed the necessity to amalgamate chemistry and physics: thus chemistry
should express its data in terms of matter and motion, in terms
of concrete, tangible chemical entities and not in abstract, a
priori undetectable principles as has been done earlier. This
programme was carried out by the eighteenth century chemists,
especially by the founders of modern chemistry A. Lavoisier and
by J. Dalton. It is to their credit that nowadays chemists
think in terms of simple substances and real entities which
preserve their individuality as corpuscles throughout many chemical vicissitudes; transmutation of elements has lain outside
the limits of chemistry since alchemy failed. It is interesting
to note that Lavoisier's experiments on a biological system,
the alcoholic fermentation of glucose, gave rise to his law of
conservation of mass.
     J.H. van 't Hoff was the chemist who (in 1875) first used
Newtonian space to describe the structure of molecule in his
"La chimie dans l'espace". A new scope was added to the corpus-
cular theory of matter: molecules do possess a spatial pattern
of relationships between various kinds of atoms. And after the
discovery of the double-helical structure of DNA and the de-
termination of the three dimensional structure of protein
molecules, nobody can deny that chemistry is in control of the
spatial molecule structure of living matter, including macro-

molecules. The expression molecular biology is sometimes used
in this respect, but the plain expression (bio)chemistry should
be preferred.

Can we go as far as A Kossel did in 1921 when he considered the chemical description of the parts of living matter as
a continuation of morphology into the molecular level and stated that in both branches of science the same method should be
used? This claim will be rejected instinctively by many workers
in the field of biology, although only a few will do it as out-
spokenly as J.H. Woodger[8] did in 1919: "The biochemist if he
is to study the parts of living organism from a purely chemical
standpoint, must, it seems, not only ignore any organization
above the chemical level, but he must also destroy it in order
to apply his methods, unless he confines himself to such fluids
as blood and urine which can be drawn from the body. The information so obtained is of the utmost importance and interest,
but it is information confined to the chemical level of organization". Although Woodger's statement is bold, it opens the
door for discussion because the form of living organisms is not
considered as sacred. Form is a manifestation of organisation
both in living material and inorganic matter, and organisation
is connected with function. This concept makes it conceivable
that the gulf between morphology and chemistry can be bridged
if we can analyse the organisation with the means of physical
science.

4. THE CELL

A keystone of the organisation in living organisms is the
cell. The importance of the cell as the basic unit of life was
fully recognised by R. Virchow[9] (1959). His theorem "Omnis
cellula a cellula", 1885, defined the continuity of living matter. This statement can be defended with lasting success as the
chemist L. Pasteur, 1867, demonstrated when he put an end to
the doctrine of Generatio Spontanea or Generatio Aequivoca that
for centuries had dominated in biology with an authority derived from Aristotle. "All life is bound to cells, and the cell
is not only the vessel of life but the living part of it", remarked Virchow; in the cell resides the Vita Propria. That the
body of the new individual is developed from a single cell,
formed by the union of the germ cells of the parents, seems
conclusive evidence in favour of this statement; moreover, the
adult body is composed of cells. Automatically cell theory was
under suspicion of vitalism. Virchow defended his Cellular Pathology against this accusation as follows: "Nothing prevents
anyone from calling such an attitude vitalism. It should not,
of course, be forgotten that no special life-force can be discovered, and that vitalism in this sense does not of necessity
signify either a spiritual or a dynamic system. But it is like-

wise necessary to understand that life is different from processes in the rest of the world, and cannot be simply reduced
to physical and chemical forces". The last sentence cannot be
considered as a convincing anti-vitalism argument.

In the cell resides the Vita Propria; but Virchow was
aware of the fact that the cell is not a vessel filled with
matter to which simple life can be attributed. For him neither
plasma nor fibrin nor protoplasm could be the source of life
and an excellent abode for an archaeus. Therefore he also
wrote: "Thus the living cell is only an autonomous part in
which known chemical substances, with their usual properties,
are ordered together in a particular manner and act in conso-
nance with their ordering and properties. This activity cannot
be other than mechanical. In vain has man attempted to find an
opposition between life and mechanism; all experience leads to
the conclusion that life is a particular kind of movement of
specific substances which, following appropriate stimuli or
impulses, act according to an inner necessity".

Not only Virchow's mind but also the general opinion in
biological science always swings back and forth like a pendulum
between frank physico-chemical conceptions of life and various,
often hidden, modifications of vitalism. At the moment the modification of Cell Biology is in vogue still, although (bio)-
chemistry has been extremely fruitful for several decades.

5. DYNAMIC AND STATIC STATE OF LIVING MATTER

(Bio)chemists were equally successful in elucidating the
"particular kind of movement" of specific substances as they
were in elucidating their molecular structure. A wealth of
knowledge concerning the pathways of intermediary metabolism
was obtained; and already in 1913 the "fundator et primus
abbas" of biochemistry in the United Kingdom, F.G. Hopkins[10]
dared to coin the epigram: "Its life is the expression of a
particular dynamic equilibrium which obtains in a polyphasic
system".

Since then our knowledge of the structure and the metabolism of the cell has grown enormously and the special role of
the cell as a keystone in the organisation and the continuity
of living matter could be linked with the molecular structure
of DNA. As far as we know most of the information necessary for
the deployment of a whole organism from a single cell is available in the DNA-structure of this cell.

A striking feature of living organisms is the contrast
between their stability, the static state of their form, and
the highly dynamic aspects of their metabolism, a stream of
chemical reactions flowing on. For a long time the structural
material and the metabolites of living organism were regarded
as basically different; endogenous and exogenous metabolism

were distinguished from each other. But the classical work of R. Schoenheimer (1940) on isotopes blurred the distinction between structural compounds and metabolites and unified endogenous and exogenous metabolism. Schoenheimer[11] stated: "The new results imply that not only the fuel but also the structural materials are in a steady state of flux".

Schoenheimer's results confirmed the idea that a living organism is an open system in a state of particular dynamic equilibrium. In the beginning of tracer experiments the attention was fixed on the dynamic state of living matter to such an extent that the wealth of information obtained by morphologists was temporarily neglected. Soon, however, the polyphasic character of living matter was taken into account and the concept of compartment became indispensable in tracer studies.

It was realised that a living organism is in a complex state of continuous renewal on the molecular and the cellular level. On the molecular level a perpetual conversion is going on, the atoms preserving their individuality throughout this metabolic process; an everlasting exchange of atoms and molecules with the environment takes place. But, although exceptions seem to occur, the tenet of DNA constancy and stability rest on firm grounds; the rate of DNA-turnover seems to be correlated only with mitotic activity. The renewal of cell populations depends on the cell type. To name two extremes no renewal is observed with neurons of all types, but epidermis cells are renewed several times during the life span of a mammal.

On the basis of this knowledge the antithesis between form and metabolism can no longer be attributed to different aspects of the molecular, cellular and supercellular structure. It is logical that the unbreakable linkage of statical and dynamical aspects of living matter introduces profound theoretical difficulties. However, this is not an exclusive phenomenon of living matter but a classical problem, already present in Greek philosophy Heraklitos said: "You cannot step twice into the same rivers, for fresh waters are ever flowing in upon you". And Greek philosophers struggled, as we do, to overcome the difficulty of explaining time-dependent phenomena with the help of time-independent static states.

Even the molecule concept is not free of this difficulty. A molecule is considered a spatial structure, not motionless but essentially with a time-independent internal arrangement. But in contrast the chemical characteristics of a molecule do at the same time imply a change in that spatial structure. In biochemistry this contrast is an important issue as these two aspects of a molecule are linked with two characteristic properties of living matter: form and metabolism. Is it possible to synthetise these two antagonistic qualities on the molecular level with the aid of quantum mechanics?

6. UNCERTAINTY RELATION

The incorporation of quantum mechanics in chemistry implies the uncertainty relation and intervenes with reducing
chemical problems to complete causal system. The usefulness of
the wave character of the electron was fully proved when the
nature of the chemical bond, and the structure and reactivity
of molecules, could be better explained although often only a
qualitative treatment of simple molecules was possible.

N. Bohr (1933)[11] paid attention to the application of
quantum theory to biological problems on the basis of the postulate that the existence of life must be considered as an elementary fact that cannot be explained. Bohr concluded we would
doubtlessly kill an animal if we pushed the investigations so
far as to be able to describe the role played by single atoms
in a vital function. In fact, he suggested an important ana-
logue may exist in biology for the Heisenberg uncertainty principle. According to this idea, full knowledge of a biological
system, although indispensable for prediction, is impossible;
and the spectre of vitalism emanates from quantum theory! But
let us dispense with additional laws for living organisms! Biochemistry is still firmly based on atomism, founded by Dalton,
as the wave length of a proton, 0,008 mm, is much smaller than
the interatomic distances. Conversion of one compound into another is explained by changes in the mutual positions of atoms.
The atoms preserve their individuality throughout the chemical
reactions; the metabolism of living matter takes place at the
molecular level. A strong argument in favour of the validity of
atomism in biological problems, and in favour of the tenet that
atoms preserve their individuality throughout many vicissitudes
in a living organism and are not affected by the processes
going on in living matter, is the fact that a link between an
aberration in atomic structure and sickness is never observed.
The possibility of a "tunnel effect" of a proton in a hydrogen
bond in DNA must be mentioned because it may lead to a deterio-
riaton of the genetic code. But the chances are so small that
the stability of the genetic code in the Watson-Crick model
does not come into question at all[13].

On the contrary molecular diseases are numerous. A minor
structural disorder, such as the difference between haemoglobin
A and haemoglobin S, can lead to a serious disease. The inborn
errors of metabolism are the best proofs for this statement. Of
course, it will be helpful to understand the important relation
between the dynamic and static aspects of a molecule, especially the important macro-molecules, if it is possible to develop
dynamic theories on long range quantum mechanical phase corre-
lations in biological systems. But this cannot be a substitute
for atomism.

The introduction of the concept or chance or probability
or uncertainty via quantum theory did not alarm chemistry; che-

mists have long since been familiar with the idea of the impossibility of being always able to predict the fate of a particle by a sequence of causes. C.M. Guldberg and P. Waage discovered, in 1865, the law of mass action governing chemical reactions. The reaction velocity v of the reaction $\Sigma v_i A_i \rightarrow \Sigma v_j P_j$ equals, according to this law,

$$v = \frac{d(A_i)}{dt} = k_1 x(A_1)^{v_1} x(A_2)^{v_2} x \ldots.$$

The reverse reaction can be treated in the same way. (A_i) denotes the concentration of compound A_i, i.e. the number of moles per litre; chemical reaction velocity is a scalar quantity. It is clear from the equation that every particle A_i has the same chance at conversion and so it is impossible to predict which of the particles will be next. The compartment with particles A_i is not only homogeneous but all particles are kinetically equal. It is an example of supreme symmetry.

Pioneer work was done in 1884 by J.H. van 't Hoff[14) (1885) who stated that chemical equilibrium, although constancy in the concentrations exists, is not a static state but a dynamic one; on the molecular level conversions take place continuously.

7. APPLICATION OF THERMODYNAMICS

The first law of thermodynamics, R. Mayer's principle of conservation of energy, was confirmed in numerous experiments with biological systems. But the second law, and in particular the theorem that a system is evolving to a state with maximum entropy, evoked strong resistance. It is immediately obvious, biologists remarked, that the tendency of living organism is to organise their surroundings (that is to produce "order" where formerly "disorder" existed) demonstrating exactly the reverse of what the second law of thermodynamics dictates! Of course, the terms order and disorder mean the presence or absence of specific patterns and they are not only based on statistics. Living organisms seem to violate classical thermodynamics; it was even suggested by H.L.M. Helmholtz that living matter could act like Maxwell's famous demon. P.W. Bridgman (1941)[15) stated correctly that "Thermodynamics recognizes no special role of the biological" and thereby broke this spell of vitalism.

I. Prigogine and F.J.M. Wiame(1946)[16), applying thermodynamics of irreversible processes, refuted the biological arguments against the second law of thermodynamics. A living organism is an open system which exchanges energy and matter with its surroundings. The positive entropy production, $d_i S$ according to the second law, which is the result of many irre-

versible, mostly metabolic processes, is counterbalanced by a
negative supply of entropy, $d_e S$. Hence dS can be zero and, if
the outflow of entropy is sufficiently large, the entropy of
the system may even decrease. It was suggested that living
organisms were in a stationary state characterised by minimum
entropy production. It is clear from this treatment that nonequilibrium thermodynamics has attractive features in dealing
with biological problems. However basic questions have still to
be answered. In the grown-up organism dynamic processes on the
molecular level are going on and nevertheless the time-
invariant condition is obvious.

But the heterogeneity of living matter is also obvious.
Anatomy, histology and electron microscopy have revealed numerous minute structures on the cellular and subcellular levels
and the morphologic evidence of their function. Biochemistry
has revealed the wealth of molecular structures in living matter and disclosed that structural material is also in a continuous dynamic state on the molecular level. And it is doubtful
that living organisms are in a near-equilibrium stationary
state, as the metabolic processes involved are mostly non-
linear.

Prigogine (1969) and his school[17] have done remarkable
work on the application of generalised thermodynamics to biological systems. They suggested that creation of supermolecular
structures may occur by non-linear kinetic processes beyond the
stability of the thermodynamic branch. This refers to symmetry
breaking instabilities as irreversible chemical reactions,
diffusion processes, and heat transport. Examples of
supermolecular structures created by dissipative processes are
the temporal oscillations appearing in the Zhabotinsky reaction and the Bénard phenomenon. Such spatial organisation and
time-dependent phenomena can be obtained from an initially homogeneous system; by the irreversible chemical processes and
diffusion processes the perfect symmetry of the compartment
defined according to the law of mass action is broken. The particles are no longer homogeneously distributed and kinetically
equal, and heterogeneity is the result. In Prigogine's theory
the assumption of "local equilibrium" is essential and he uses
the concept of "department" in his kinetic considerations.
Therefore it seems useful to pay attention to the picture produced by tracer studies of homogeneity versus heterogeneity in
living matter.

8. HOMOGENEITY AND HETEROGENETY IN LIVING MATTER

The first tracer studies, by Schoenheimer on the dynamic
aspects of living matter, were rather qualitative in nature.
They proved the rapidity with which molecules exchange. Quantification requires a careful definition of the concepts used and

the problems studied. The pool is the total amount of a substance in a system or a subsystem; for instance the calcium
pool in plasma or the iron-ferritin pool in the mitochondria,
or in the pyruvate pool in muscle. The concept of "compartment"
lies at the heart of the tracer study of metabolism in a living
organism. On the one side it is the expression of the heterogeneity so characteristic for living matter. On the other side a
compartment is a homogeneous metabolic pool, where not only the
concept of concentration can be used but where the particles
are kinetically equal. A compartment in tracer studies is identical with a compartment as used in the law of mass action. The
cardinal question is: do such compartments, or such homogeneous
metabolic pools exist in living matter? Or does heterogeneity
persist up to the molecular level?

Experiments with D_2O and Tr_2O made it clear that no bar-
riers exist for H_2O molecules in living organisms. Furthermore,
it has been proved that mixing within the total body water pool
in man is completed within five hours. As the total body water
amounts to 2200 mol or 40 litre in average man and is normally
renewed with a velocity of 110 mol or 2 litre per day, it seems
reasonable to consider total body water as a compartment. It
can be proved that for a number of compounds, such as urea, the
same is true. But for a large number of solutes the cell
membrane is an obstacle dividing the total body water into an
extra-cellular and an intra-cellular space. Many compounds such
as electrolytes, glucose, aminoacids etc. are extra-cellularly
present in compartments, but the intra-cellular situation is
much more complex. Even a homogeneous metabolic pool of aminoacids as a source for protein synthesis does not seem to exist.
The intra-cellular aminoacid pool is subdivided in relation to
the different organelles of the cells; and in protein turnover
several situations can be distinguished. There is continuous
and random turnover; the molecules have no history. But a
haemoglobin molecule has a finite life span of 120 days, equal
to the erythrocyt life-span; and not only does a time structure
exist intra-cellularly, but also important enzymes systems
within the cell are spatially co-ordinated. Of course, the DNA
molecule has an extraordinary place in the renewal process:
there is no renewal unless mitosis including duplication of DNA
occurs.

It is evident that the constancy of extra- and intra-cellular parts of the water department in an organism is a vital
condition. As H_2O molecules can freely move throughout the
whole water compartment it is clear that the maintenance of the
constancy of extra- and intra-cellular water is an osmotic
problem, and an equilibrium must exist. The constancy of the
extra- and intra-cellular volumes is maintained by keeping the
osmolality of both parts constant and equal. The main point of
this regulation is the maintenance of the osmolality of the
extra-cellular part at about 320 mmol/kg H_2O. This regulation

comprises many and various dynamic processes, not all of them
explicitly described as yet. The maintenance of the constancy
of extra-cellular fluid osmolality is only a part of the very
complex system maintaining the overall constancy of many variables in the extra-cellular space. W.B. Cannon coined the term
homeostasis for this regulator system. Obviously he did not
take it for granted that homeostasis is based on physico-chemical principles as he titled a famous book "The Wisdom of the
Body". The significance of the constancy of the extra-cellular
fluid was formulated by C. Bernard in a concise way: "La fixité
du milieu interieur est la condition de la vie libre". Numerous
studies have been done on the nature of homeostasis, but our
insight is still limited.
      The circadian rythm of water and electrolyte in man gives
an inkling of the oscillatory nature of homeostatic processes.
It is interesting to see the impacts on the day and night
rhythm of water and electrolyte excretion when a subject is
forced to live a 27 hour day, for instance, instead of a 24
hour day, and thus eight artificial days instead of nine real
days. Water excretion shows eight peaks and has thus adapted to
the artificial time according to which the subject was living,
whereas potassium shows nine excretory peaks, occurring around

Table 1. Molality of electrolyte content of the extra- and the
         intra-cellular fluid

	Extra-cellular	Intra-cellular
Sodium	152	15
Potassium	5.5	150
Calcium	5.5	2
Magnesium	2	27
Bicarbonate	29	10
Chloride	110	1
Sulfate	1	–
Organic acids	5.5	20
Proteinate	17	63
Phosphate	2	100

midday by real time. This suggests the continued operation of a
24-hour clock from which water excretion has escaped under the
influence of a 27-hour cycle. A bifurcation approach seems
worthwhile but our lack of knowledge of the mechanisms is a
stumbling block.

9. THE CHEMICAL BASIS OF MORPHOLOGY

The great differences (see Table 1) between the content of the extra- and the intra-cellular fluid are maintained by the metabolic, energy-providing, processes going on inside the cells. It is tempting to apply thermodynamics of irreversible processes to this coupling of transport and chemical reaction using the coupling coefficient in linear phenomenological equations on the base of an anisotropic membrane. A. Katchalsky and O. Kedem[18] have done important work in this field; and B.K. van Kreel[19] in our department has studied placenta transport.

But a fundamental difficulty arises: the metabolic processes responsible for the energy production are nearly all catalysed by enzymes and have non-linear kinetic equations. Moreover, since the classical work of H.A. Krebs, 1932[20], it is evident that many metabolic processes in living matter are cyclical reactions. The famous chemical principle of detailed balancing, an early precursor of Onsager's principle of microscopic reversibility, is not valid. Consequently, although a degree of homogeneity exists, a living organism is not in a stationary state near equilibrium with a minimum entropy production.

In the classical physiological approach of the osmotic equilibria between the intra- and the extra-cellular fluid, and in the approach with thermodynamics of the near-equilibrium state, the existence of supermolecular structures like membranes is taken for granted. Prigogine's approach to membranes and similar supermolecular structures is revolutionary as it starts with a uniform state and assumes the creation of inhomogeneities, which are the result of the non-linearity of the kinetic processes and can act as biological structures.

A.M. Turing(1952)[21] was the first to suggest such a chemical basis of morphogenesis. It is very attractive to associate biological structures on a supermolecular level with chemical instabilities that lead to the spontaneous self-organisation of a biological system. From the tracer studies it follows that homogeneity exists in living organism in the extracellular space, but also to a certain degree inside the cell. Tracer studies have also revealed that even the structural material of a living organism is in a continuous state of renewal on the molecular level. Therefore it is conceivable that dissipative structures are continuously created in living matter as a consequence of the non-linear chemical and transport processes in metabolism according to Prigogine's theory.

But the dissipative structures observed in the case of the Zhabotinsky reaction and the Bénard phenomenon, impressive as they are, disappear in the equilibrium state; whereas classical biological and morphological structures, for instance membranes, still remain when transport and chemical processes come to an end. How should one deal theoretically with this discre-

pancy between Prigogine's theory of spontaneous self-organisation in a biological system initiated by chemical instabilities
and the highly ordered structure already present? The work of
J.P. Changeux and others(1967)[23] on the co-operativity of biological membranes indicated that a superposition of new supermolecular structures created by dissipative processes takes
place on the basis of old established morphological structures.
These morphological structures are mostly integral parts of the
cells and the question of how these structures came to be created remains to be answered. As far as we know the construction
of a cell is only possible with the information available in
the DNA-molecule. This biological memory is assumed to be renewed (i.e. duplicated) only in mitosis. It is responsible for
the coherence of an organism as a whole and for the individual
form. This biological memory is an accumulation of genetic records gained during the evolutionary process. The single DNA-
molecule of E. Coli has a relative molecular mass of 2×10^9
and an extended length of almost 1 millimeter; in man these
figures are a tenfold higher. The deployment of a living organism seems to be a perfect performance of a play created and
developed by trial and error during its evolution. It does not
seem to be a spontaneous self-organisation at all; it is mostly
determinism, as in the picture which J. Needham included in his
book. Bifurcation is a restriction on the possibilities. Only a
mistake of a pointsman can lead to an unexpected situation, and
this is in full accordance with the aphorism: "Nullis Morbus
Sine Defecta - no disease without a defect". Molecular diseases, inborn errors of metabolism, are exemplary models. Of
course, the environment and chance can influence the development of a living organism, as is done in modelling the development of the brain by the learning processes. It is also conceivable that pathology is caused by perturbation of an intact
physiological control system of a grown-up organism. But it
cannot be denied that a living organism owes much to determinism personified by DNA-structure and the consequent exclusion
of chance. Of course, the process of spontaneous creation of
supermolecular structures may be of great significance in the
evolutionary process and the pre-biological stage. Besides,
this process is brought under strict control by the molecular
biological memory, the DNA-structure, which is mutated or extended or kept constant by the interactions of necessity and
chance, according to Democritus' famous postulate; everything
happens by chance and necessity. Prigogine's theory is an
important step forward in our understanding of supermolecular
structures in living matter. But the fact that living matter do
keep a molecular record governing the organism must be kept in
mind.

NOTES AND REFERENCES

1. See P. Glansdorff and I. Prigogine: 1971, Thermodynamic theory of structure, stability and fluctuations.

2. See O. Gurel and O.E. Rössler: 1979, Bifurcation theory and applications in scientific disciplines, Annales of the New York Academy of Sciences.

3. Needham, J.: 1936, Order and Life.

4. Haldane, J.S.: 1932, Order and Life.

5. Brown, R.: 1828, Phil. Magazine 4, p. 161. Ann. d. Phys. u. Chem.: 1828, no. 14, p. 294.

6. Einstein, A.: 1926, Investigations on the theory of the Brownian movement, see also the notes by R. Fürth, Dover reprint 1956.

7. Bernard, C.: 1878, Leçons sur les phénomènes de la vie communs aux animaux et aux végétaux.

8. Virchow, R.: 1959, Disease, Life and Man, Selected Essays by Lelland J. Rather.

10. Needham, J. and Baldwin, E.: 1949, Hopkins and Biochemistry.

11. Schoenheimer, R.: 1941, The Dynamic State of Body Constituents.

12. Bohr, N.: 1933, Light and Life, Nature 131, pp. 421-457.

13. Marios, M.: 1969, From Theoretical Physics to Biology, Proc. of the Second Int. Conference on Theoretical Physics and Biology, see p. 240 etc.

14. Hoff, J.H. van t': 1885, Etudes de dynamique chimique.

15. Bridgman, P.W.: 1941, The Nature of Thermodynamics.

16. Prigogine, I. and Wiame, J.M.: 1946, Experientia 2, p.451.

17. Prigogine, I. and Glansdorff, P.: 1973, Bull. Acad. Roy. Belg. Cl. Sci. 59, p. 672.

18. Katchalsky, A. and Curran, P.F.: 1965, Nonequilibrium thermodynamics in biophysics.

19. Kreel, B.K. van: 1981, A mathematical approach to
    mechanisms of placental transfer, contribution to: <u>Transfer
    across the Primate and Non-primate Placenta</u>, edited by
    H.C.S. Wallenburg et all.

20. Krebs, H.A.: 1946, <u>Enzymologia.</u> Vol. XII, 2, p. 88.

21. Turing, A.M.: 1952, <u>Phil. Trans. Roy. Soc. Ser. B</u>, V. 237,
    p. 37.

22. Changeux, J.P., Thiéry, J., Tung, Y. and Kittel, C.: 1967,
    <u>Proc. Nat. Acad. Sci.</u> 57, p. 337.

THE ANALYSIS OF BIFURCATION PHENOMENA ASSOCIATED WITH THE
EVOLUTION OF URBAN SPATIAL STRUCTURE

M. Clarke and A.G. Wilson

School of Geography
University of Leeds

1. INTRODUCTION

One of the more interesting developments in urban and
regional modelling in recent years has been the analysis of the
bifurcation properties of a wide variety of models. Just as Jay
Forrester's (1969) attempts to explicitly focus on urban dynamics provided the impetus for those within the modelling discipline to move away from purely static analysis, the innovative
work of Thom, Zeeman and others has brought the notion of
bifurcation behaviour to the notice of those currently engaged
in urban modelling. In this paper we outline how we have
attempted to analyse the bifurcation properties of a well-known
set of spatial interaction and activity models, both in their
equilibrium and disequilibrium formulations. In addition, we
give a number of illustrations of numerical simulations performed for both a 729-zone hypothetical system and also a 30 x
900 Leeds system where we explicitly focus on the problem of
modelling urban spatial structure. (A much fuller set of
results can be found in Clarke and Wilson, 1984).
In the social sciences in general and urban modelling in
particular, the use of methods based on catastrophe theory, and
to a lesser extent bifurcation theory, have, on occasion, met
with a hostile reception. We attempt to outline reasons for
this in the next section.

2. BIFURCATION THEORY IN THE SOCIAL SCIENCES

Many of the social sciences, particularly economics, geo-

M. Hazewinkel et al. (eds.), Bifurcation Analysis, 67–99.
© *1985 by D. Reidel Publishing Company.*

graphy and psychology, have experienced substantial attempts at
the quantification of socio-economic relationships. This has
taken both a statistical and mathematical form and it is the
latter that we focus on in this paper. Much of this quantification has borrowed heavily from analogies in the physical
sciences, for example the concept of an economic equilibrium
can be seen to be derived from the notion of chemical and thermodynamic equilibria in chemistry and physics, and in geography
the gravity model was originally formulated directly from
Newton's gravitational laws (see Ravenstein, 1889). It comes as
no surprise, therefore, that when certain techniques or methods
ascend to the heights of popularity in the natural sciences
that the social scientist is alerted to the possibility of
application in his or her discipline. This, of course, does not
necessarily imply that there is a one-way flow of methods from
the natural to the social sciences, indeed, some techniques,
such as principal components analysis, were as much the responsibility of psychologists as statisticians.

However, in the case of bifurcation theory, and in particular catastrophe theory, there has been considerable interest
and a growing number of applications in the social sciences
(see Wilson, 1981, for a review). In parallel with this work
have been several criticisms of the use of these methods in the
social sciences (e.g. Zahler and Sussman, 1978). We wish to
demonstrate that while much of this criticism is justified,
this is more a result of the type and nature of some of the
applications than the appropriateness of the use of these
methods in the social sciences. Three different uses of the
findings from bifurcation and catastrophe theory are now identified: the first is where the manifolds derived from catastrophe theory, more than often the cusp surface, are used in a
descriptive or qualitative way; the second is where some aspect
of human behaviour is described as a potential function of the
canonical form of one of the seven elementary catastrophes;
finally there are those applications which analyse the bifurcation properties of a much wider class of non-linear models of
socio-economic systems. It is within this latter category that
our own efforts, described in the subsequent sections of this
paper, fall.

The use of catastrophe theory to describe certain types of
dynamic behaviour has proved extremely attractive to many
social scientists. The appeal of this approach probably lies in
the way that a variety of very different system dynamics can be
described without the necessity of constructing a formal mathematical model of the system. In Figure 1 the cusp catastrophe
is illustrated. Most applications consider two alternative
states of the system, denoted here as X_1 and X_2 and two control
parameters, a and b. Changes between the two states, X_1 and X_2,
can then be 'explained' in terms of the different combinations
of parameter values, a,b. While many of these examples can be

commended for their novelty most of them clearly start with the
cusp catastrophe in mind and then attempt to fit some socio-
economic phenomenon to the surface, often with little evidence
of justification for doing so.

The next set of examples go one step further. They begin
with the potential function of one of the elementary catastrophes (again usually the cusp) and attempt to justify that

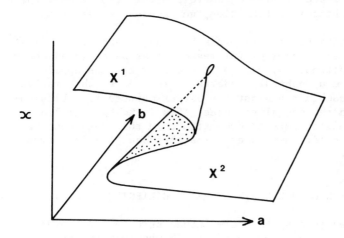

Figure 1.

this is the appropriate function to represent some type of
behaviour (often a choice process). So, for example, the equation given by (2.1) where x is the state variable and a and b
are parameters has been posited, amongst other things, as a representation of the utility of travel (Dendrinos and Zahavi,
1980), an urban welfare function (Isard and Liossatos, 1977),
and population density in an area of Greece between the second
and seventeenth century (Wagstaff, 1978).

$$Z = \tfrac{1}{4}x^4 + \tfrac{1}{2}ax^2 + bx \qquad\qquad (2.1)$$

While there is good reason to believe that a polynominal
function represents many different types of behaviour, it does
seem that some heroic assumptions are being made about social
systems so that this canonical form of the cusp catastrophe can
be so conveniently derived. It comes as no real surprise when,
after these models have been formulated, the systems they pur-
portedly represent show all the properties of the cusp catastrophe, such as jumps, hysteresis, and so on.

The third category of applications are of a rather different nature. They take as their starting point some model of

human activity or behaviour. If this model has non-linearities
then bifurcation theory informs us there is the possibility of
discovering certain conditions when the nature of the model
solution changes in response to small changes in the value of
the model's parameters or exogenous data, so-called bifurcation
points. Whereas this type of behaviour often bears strong resemblance to catastrophe theory-like behaviour, it does not
rely on catastrophe theory whatsoever. Bifurcation theory has a
much older mathematical tradition, going back at least to
Poincare.

The majority of criticism has been addressed at the first
two of the above categories for reasons that have already been
sketched. Unfortunately, many of these over-zealous applications that were undoubtedly undertaken in good faith have lead
to a general backlash against the use of methods from bifurcation theory in the social sciences. This is to be regretted,
for, as we shall attempt to demonstrate in the subsequent sections of this paper, many useful insights can be gained from
these methods.

3. THE EVOLUTION OF URBAN STRUCTURE: EQUILIBRIUM ANALYSIS

What do we mean by the term 'urban structure'? Broadly
speaking, cities are composed of people who perform activities,
such as employment, travel, shopping, recreation. The physical
entities in which these activities take place, such as factor-
ies, shops, and residences are the structures which we are
interested in modelling. More specifically, we aim to develop
models that examine how the size and location of urban structures changes over time in response to a number of factors. To
illustrate the general problem we use the example of retail
facility location but argue that the methods that are developed
are applicable to a wider set of problems.

The traditional use of retail models in planning has been
to predict the flows of revenue into a zone given a certain
distribution of retail supply. The planning function was
commonly to assess the impact of additions to the retail stock
on the revenues of other centres. Often this involved comparing
the 'before' and 'after' situation and on the basis of the
forecasts of the impact of the proposed development, planning
permission was either granted or denied.

The model most often used for these purposes was based on
the well-known Huff (1964) and Lakshmanan and Hansen (1965)
shopping model

$$S_{ij} = A_i e_i P_i W_j^\alpha e^{-\beta c_{ij}} \tag{3.1}$$

where

$$A_i = \left[\sum_j W_j^{\alpha} e^{-\beta c_{ij}} \right]^{-1}$$ (3.2)

to ensure

$$\sum_j S_{ij} = e_i P_i$$ (3.3)

and S_{ij} is the flow of revenue from households in i to shops in j, e_i is per capita expenditure on retail goods in zone i; P_i is the population of zone i; W_j is the amount of retail supply in zone j; measured, say, in square metres of retail floorspace (and is taken as a measure of attractiveness); c_{ij} is the cost matrix; α is a measure of consumer scale economies; and β is a parameter measuring the elasticity of travel demand.

In the work described here we take a fundamentally different approach to the retail problem. We are interested in the modelling of the structural variables, which we take to be the W_j's. So for a given system and parameter values we wish to determine the values of $\{W_j\}$. This can be achieved according to a number of hypotheses, which it can be shown turn out to be equivalent. These range from consumer surplus maximisation, producer profit maximisation to revenue-cost balancing (see Clarke, 1982). Here we use the revenue-cost balancing apporach based on the findings presented in Harris and Wilson (1978), subsequently modified by Wilson and Clarke (1979) and Clarke (1984).

The Harris and Wilson argument was based on the assumption that the suppliers or producers of retail facilities will expand if their revenue exceeds their costs and will contract if costs exceed revenue. If we define k as the unit cost of providing retail floorspace and D_j as the revenue acruing to zone j then we can write these conditions as:
if

$$(D_j - kW_j) > 0 \qquad\qquad W_j \text{ will expand}$$ (3.4)

and if

$$(D_j - kW_j) < 0 \qquad\qquad W_j \text{ will contract}$$ (3.5)

The equilibrium conditions are considered to be when revenue exactly balances costs. That is, when

$$D_j = kW_j$$ (3.6)

As $D_j = \Sigma_i S_{ij}$ we know from equation (3.1) that

$$D_j = \Sigma_i \frac{e_i P_i W_j^\alpha e^{-\beta c_{ij}}}{\Sigma_k W_k^\alpha e^{-\beta c_{ik}}} \qquad (3.7)$$

and by substitution that

$$kW_j = \Sigma_i \frac{e_i P_i W_j^\alpha e^{-\beta c_{ij}}}{\Sigma_k W_k^\alpha e^{-\beta c_{ik}}} \qquad (3.8)$$

It is immediately clear that these equations are highly non-linear in the W_j's and prove impossible to solve analytically. Another notable feature is that the revenue in any one zone is a function of the values of W_j in all the other zones, so equation (3.7) and (3.8) can be considered as competitive models where producers each compete for available revenue. It is this complex nature of these models that gives rise to much of the interesting dynamical behaviour we discuss shortly. Let it also be noted that these complexities are also responsible for many of the difficulties we have in the numerical simulations and in analysing the mechanism which give rise to change. We now show how this model is equivalent to the consumer surplus problem discussed in the preceding section.

The next step in the Harris-Wilson argument was based on the relationship between the D_j-W_j curve and the equilibrium condition $D_j = kW_j$, which is clearly a straight line of slope, k, on the same axes. Thus the intersection of the D_j-W_j curve and this straight line would identify potential equilibrium points for a number of different cases.

Three different cases were identified and we discuss each of these in turn.

(i) $\alpha < 1$ (Figure 2)
   Here two equilibrium points are identified: W_j^A which has a positive value and is stable and $W_j = 0$ which is unstable. This implies that if the retail model is run with $\alpha < 1$ each zone will have a non-zero W_j.
(ii) $\alpha = 1$ (Figure 3)
   In this case the D_j-W_j curve does not necessarily intersect the straight line and there is the potential for W_j to be

either zero or non-zero, with the equilibrium point now stable.

(iii) $\alpha > 1$ (Figure 4)

Now the qualitative nature of the D_j-W_j curve changes with the emergence of a point of inflection, and we have three equilibrium points depicted, W_j^A, W_j^B and $W_j=0$. The upper equilibrium point, W_j^A, is stable as is the lower equilibrium point $W_j=0$. W_j^B is unstable.

Harris and Wilson then examined how a particular zone could potentially move from the W_j^A equilibrium point to $W_j=0$ (or vice versa), under the assumption that all the other variables (including the W_k's ($k\neq j$)) remain fixed. The mechanism they proposed was that the slope of the $D_j=kW_j$ line could

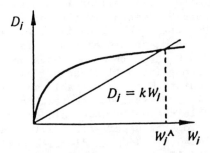

Figure 2. Revenue Curve and $D_j = kW_j$ line $\alpha < 1$
Equilibrium points W_j^A and $W_j \cong 0$

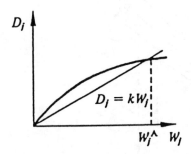

Figure 3. Revenue Curve and $D_j = kW_j$ line $\alpha = 1$
Equilibrium points W_j^A and $W_j \cong 0$

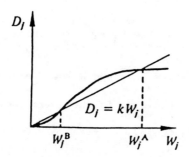

Figure 4. Revenue Curve and $D_j = kW_j$ line $\alpha > 1$
Equilibrium points W_j^A, W_j^B and $W_j = 0$

change. This would be effected by changes in suppliers unit
costs, k; if this increased then the slope of the straight line
would clearly increase. Again, assuming that everything else
remains constant the value of W_j^A will fall as k increases, and
this is demonstrated in Figure 5. Eventually a point will be
reached in the $\alpha > 1$ case, when the D_j–kW_j line is tangential to
the D_j–W_j curve (in fact W_j^A and W_j^B coalesce). The value that k
takes at this point is called k_{jcrit}. If k increases only very
slightly the upper equilibrium point will disappear leaving
only the lower equilibrium point, $W_j = 0$ as a stable solution. If
we plot W_j against k for this process then the resultant set of
curves can be seen to resemble the well-known fold catastrophe
(see Figure 6). The 'jump' from the upper stable solution W_j^A
to the lower stable point, $W_j = 0$, is seen as discrete and is a
consequence of only a very small change in k. However, as we
pointed out in the preceding section, these results are derived
independently of catastrophe theory and are a result of the
model having multiple equilibria.

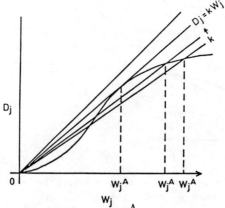

Figure 5. Changing W_j^A as a result of k-change

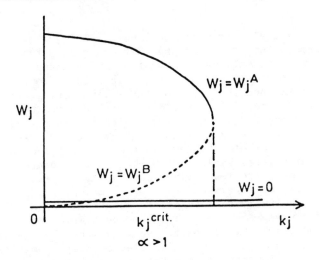

Figure 6. W_j - k Plot

A feature of the Harris-Wilson argument which generates
Figures 2 - 6 is that the form of the D_j-W_j relationship for a
particular zone, j, is deduced on an 'other' terms remaining
constant' basis. This is not the case in practice, particularly
as the total amount of stock to be allocated, W, is a constant,
and thus changes in the value of W_j in one zone will automatically force changes in the values of the other W_k's. Hence the
form of the revenue curve

$$D_j = \Sigma_i \frac{e_i \, P_i \, W_j^\alpha \, e^{-\beta c_{ij}}}{\Sigma_k \, W_k^\alpha \, e^{-\beta c_{ik}}} \tag{3.9}$$

will obviously be affected.

The relaxing of the Harris-Wilson assumption causes considerable difficulties in numerical work that attempts to construct Figures 2 - 6 from model runs. Four possibilities were
outlined in Wilson and Clarke (1979), and we refer the
interested reader to that paper for further details.

It turns out that the Harris-Wilson argument remains valid
in the general sense that the relationship between the two
curves is the mechanism which produces change (both smooth and
discrete) in the values of the structural variables. However,
it is not usually the case that changes in k will produce
bifurcation points. Instead, the interesting types of change
are produced by variation in the parameters β and α. These are
now described in turn.

(a) β-changes

β, as we have already mentioned, measures 'ease of travel' for the consumer. The effect of smoothly changing β from a high value to a low value will be to move from a dispersed pattern of centres to a centralised one. At the individual zone level β will clearly influence the shape of the revenue curve, and this is apparent in Figures 7 (a and and b). With β high the 'pick up' rate of revenue $(\partial D_j)/\partial W_j$ is very quick but soon tails off. For low β the rate is much more constant over the W_j range. This implies that for high β there is a much greater chance of the revenue curve intersecting the D_j-kW_j curve — hence the probability of many centres having small values of W_j. For low β the probability of intersection is much smaller but if it does occur it is likely to result in a relatively high value of W_j - but, of course, there are likely to be fewer centres. In the Wilson and Clarke (1979) paper, we were inconclusive as to whether a β-change itself would give rise to jumps. However, further numerical investigation demonstrated that β is respon- sible for discrete change and the mechanism is nothing more than the reverse of the behaviour under k variation. Simply stated if the D_j-kW_j line remains fixed, different values of β will generate curves with, everything else being equal, differ-

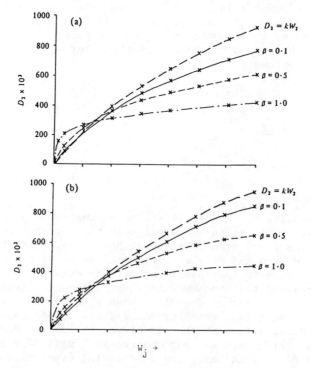

Figure 7. $D_j = W_j$ curves derived under Assumption 1
W = 1,700,000, changing α and β

ent upper equilibrium points W_j^A. There will exist a value of β,
however, which will produce an equilibrium point which is tangential to the $D_j = kW_j$ line – we can call this β_j^{crit}. Any continued change in β will result in a disappearance of W_j^A and
thus will experience a discrete change in the value of W_j. However, the way in which W_j changes for variations in β is quite
interesting. Let us assume that we start with a high β, β^I and
progressively reduce its value. Figure 8 demonstrates what
happens in the $\alpha > 1$ case.

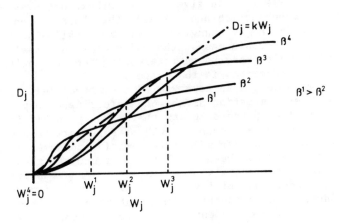

Figure 8.

For high β, W_j takes a low value. As β decreases, the value
of W_j increases because the shape of the D_j-W_j curve changes too.
Finally, the D_j-W_j curve no longer intersects the D_j-kW_j line and
the value of W_j 'jumps' to zero. If we plot W_j against β, a rather
different plot to that of Figure 6 may emerge. This is shown in
Figure 9. We present examples of this type of behaviour shortly.
Of course, this argument is again based on an 'everything else
being equal' assumption, and the behaviour described above may be
obscured by or importantly affected by behaviour in other zones.

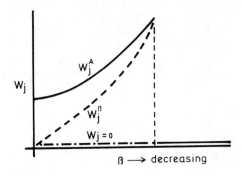

Figure 9.

(b) α-changes

At first it was thought that α acted in a fairly straight-forward, albeit important way. As Wilson (1981) pointed out $\alpha=1$ is clearly a 'critical' value. When $\alpha<1$ the $W_j=0$ equilibrium state is unstable and therefore all W_j transitions will be from and to positive non-zero values and they will be smooth. In the case of $\alpha>1$ discrete change, involving the $W_j=0$ equilibrium state is possible.

However, more recently it has been demonstrated by Clarke (1981) that α-variation can give rise to other, not previously anticipated, methods of change. Clearly the shape of the D_j-W_j curve can be affected by changing α. Clarke (1981) also demonstrated that, in certain cases, the value of k was itself a function of α. This leads to an interesting set of results which are fully described in that paper.

How do we expect α to have varied over time? This is perhaps not as easy a question as the one referring to β variation. Empirical evidence suggests that α has probably increased; the agglomeration and growth of supermarkets and hypermarkets suggests that producers perceive consumer scale economies by providing these facilities. Of course, these trends may reflect producers' scale economies and consumers' changing attitudes and ability to travel. The most important factor to determine is whether α is greater than or less than one. This is clearly dependent on the type of good being considered (and is also an aggregation problem). An attempt to analyse some of these problems can be found in Harris, Choukroun and Wilson (1981).

There are other potential mechnism for change that we need to examine if a full consideration of system dynamics is to be undertaken. Any variable on the right hand side of equation (3.7) can affect the shape of the D_j-W_j curve. Therefore, we can anticipate that changes in zonal populations, per capita expenditure and travel costs can also give rise to bifurcations, and we demonstrate these effects in the numerical simulations we describe shortly.

Because of the difficulties encountered in the above analysis, alternative methods of determining zonal criticality have been investigated. These include an examination of the properties of the determinant of the Hessian matrix of the Mathematical Programming model which is equivalent to equations (3.6) and (3.7). Another method involved the analysis of the above graphical argument in terms of calculating the derivatives

$$\frac{\delta D_j}{\delta W_j} = \Sigma_i \frac{\alpha S_{ij}}{W_j} \left[1 - \frac{S_{ij}}{e_i P_i} \right] \qquad (3.10)$$

and comparing this value, evaluated for successive equilibrium values of W_j, with the value of k. As we approach a bifurcation point these two values should tend to converge, until at the critical point they should be equal. Both these methods are described in further detail in Clarke (1984).

We now proceed to present a set of numerical simulations that illustrate some of these theoretical ideas.

4. NUMERICAL EXPERIMENTS: THE EQUILIBRIUM MODEL

In both this section and section 5 we present results derived from a large number of numerical simulations. We can in this paper only present a taste of the variety of possible experiments and results that we are able to perform. A more detailed presentation is available elsewhere (Clarke and Wilson, 1984). The organisation of these experiments has in itself posed some interesting organisational questions that are discussed in Clarke and Wilson (1981). We begin by describing the spatial systems and data assumptions used in simulations.

The first spatial system we have used is based on a hypothetical, symmetrical system. This system was constructed on a regular grid basis with a 27 x 27 array of points giving 729 zones in all. Within this grid an approximate circle is defined (see Figure 10) delimiting zones considered endogenous to the system, so giving us a high degree of symmetry. Euclidean distance was used for calculating the cost matrix. That is, given spatial coordinates of two points, x_i, y_i, and x_j, y_j the distance between them can readily be computed as

$$c_{ij} = \sqrt{(x_i - x_j)^2 + (y_i - y_j)^2} \qquad (4.1)$$

The second spatial system is based on a representation of the Leeds city system with 30 demand points, corresponding to 1971 census ward centroids, and 900 potential supply points, based on a 30 x 30 regular grid (see Figure 11). Euclidean distances are once again used for calculating the cost matrix, but clearly alternative ways of generating this data could be used.

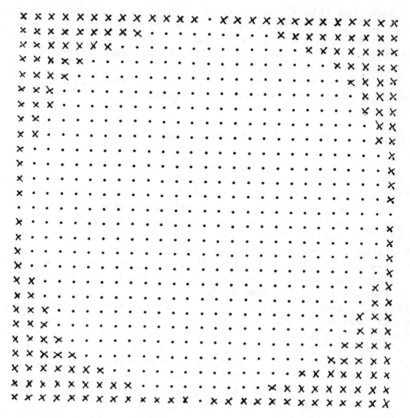

Figure 10. 729 zone spatial system. x indicates exogenous zone

The values of e_i and P_i are either taken from census data (in the Leeds case) or are assumed to be uniform or varying according to some distance decay function (in the 729 zone case).

The majority of results presented in this paper consists of computer graphical output. The x and y dimensions correspond to the spatial system, and the Z dimension represents the amount of retail facilties present in the zone (the graphical package used is called GINOSURF). Note that due to the inter-polation procedure of GINOSURF the W_j variables appear as peaks, not as histograms as we would wish.

The first set of plots examine the effect of α and β variation for the 729 zone system. Figure 12 consists of 11 model results for different combinations of α and β, around a

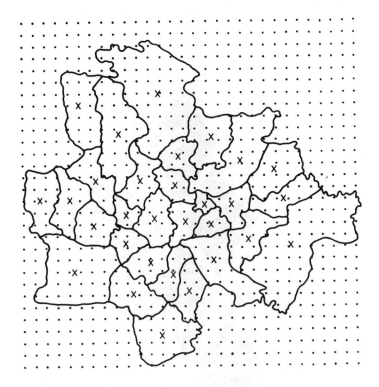

Figure 11. Leeds - 30 demand 900 supply points

pivotal case of $\alpha=1.3$, $\beta=2.5$. The results are entirely as anti-cipated in the previous section, but the variety of different patterns that emerge is interesting.

Another feature of these plots can be seen if we carefully examine the pattern variation between $\beta=2.0$ and $\beta=0.5$. When $\beta=2.0$ a facility exists in the central area. However, this dis-appears in the $\beta=1.5$ case, only to reappear when $\beta=1.0$. This emphasises that the change in the pattern of retail facility location under parameter variation is not smooth but involves discrete change. The results generated in this figure have all been derived under a uniform plain assumption, with the values of all the $e_i P_i$'s equal.

The next set of results presented in Figure 13 use an assumption about spatially varying supply costs, k_j. We assume the following relationship:

$$k_j = \frac{1}{[(c_j, 365)]^\mu} \qquad (4.2)$$

M. CLARKE AND A. G. WILSON

Figure 12. Retail structure patterns for various α and β values

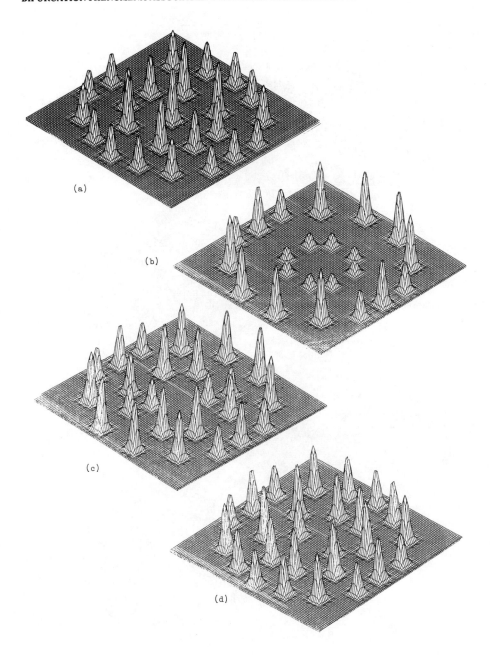

Figure 13. k_j variation

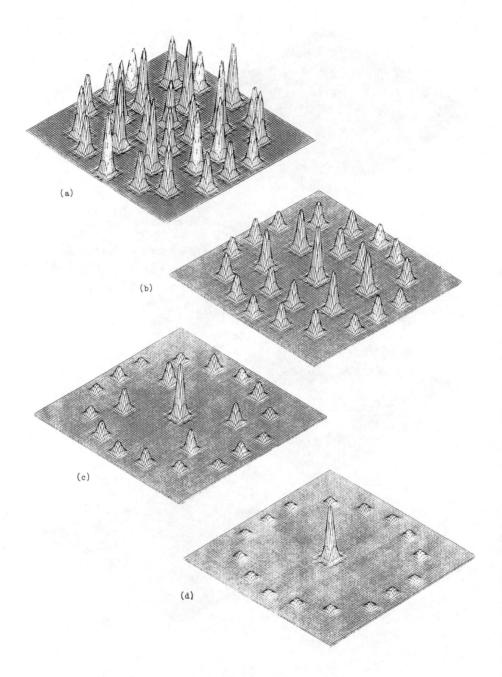

Figure 14. Cheap travel to city centre

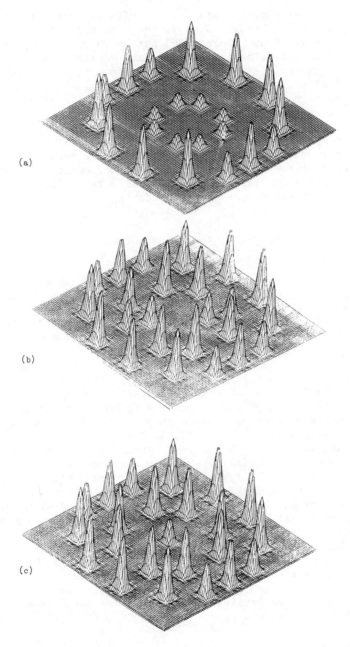

Figure 15. Combination of cheap travel and k_j variation

which means that supply costs decline with distance from the
centre (zone 365). Figure 13(a) is the plot for $\alpha=1.5$, $\beta=2.5$
without k_j variation. Figure 13(b) has $\mu=1.0$ and this produces
a suburban-dominated pattern of facility location. With $\mu=0.5$
(Figure 13(c)) this effect is reduced somewhat, even more so in
Figure 13(d) with $\mu=0.25$ with more 'central' features emerging,
but the pattern never gets back to the Figure 13a situation of
a dominant centre.

The next step in the argument is to recognise that most
urban transport systems are designed to favour travel to the
centre, through some radial orientation of the network, although this picture is complicated by higher congestion levels
near the centre. To incorporate these effects we firstly modify
the cost of travel to the centre by taking

$$c(_j, 365) = \upsilon \times \text{euclidean distance} \tag{4.3}$$

where υ takes a series of values less than one, so that the
centre is favoured to varying degrees. With $\upsilon=1$, $\alpha=1.3$ and
$\beta=2.5$ we note that no central facilities exist (Figure 14(a)).
In Figure 14(b), with $\upsilon=0.95$ there is now a central development. This intensifies considerably with $\upsilon=0.85$ and 0.75
respectively in plots (c) and (d).

Congestion is a more difficult problem to handle and we
refer the reader to a paper where we explicitly focussed on
this problem (Beaumont, Clarke and Wilson, 1981a).

An interesting effect to analyse is what happens to the
spatial patterns when we combine cheap travel to the centre
with spatially varying costs, k_j, as the effects of these two
factors are in opposite directions, one favouring suburban
centres, the other central ones. This was done to generate the
results in Figure 15. Figure 15(a) was produced with spatially
varying k_j's but no cheap travel. In Figure 15(b) we have the
same k_j variation, but with central costs factored by 0.95, but
the k_j effect still dominates. However, with $\mu=0.75$ (Figure
15(c)) we do get some central development, which demonstrates a
pattern bifurcation between 0.95 and 0.75.

The assumption concerning uniform $e_i P_i$ patterns clearly
needs to be relaxed. Broadly speaking, densities decline with
distance from the centre and per capita incomes increase. To
incorporate a broad decline of $\{e_i P_i\}$ we use the following formula for the 729 zone system.

$$e_i P_i = a \times e^{-bc_i, 365} \tag{4.4}$$

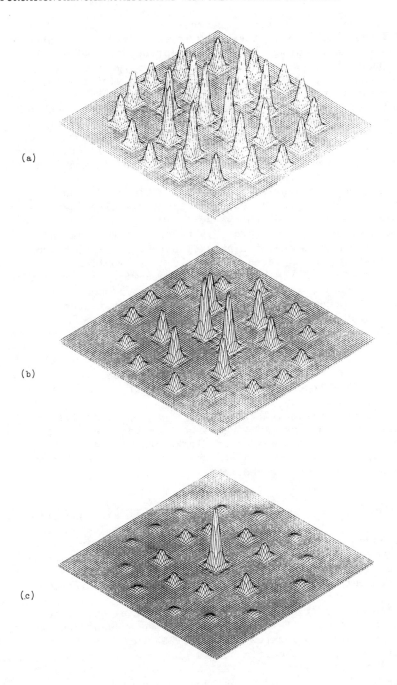

Figure 16. $e_i P_i$ variation

where a is taken as 500 and b, successively in Figure 16, as
0.1, 0.3 and 0.5, thus representing increasingly steep rates of
decline. In addition, we create a population "hollow" in the
central zone, because in most cities few people reside in the
very central area; this is achieved by multiplying the central
zone population by 0.5. The three plots shown in Figure 16 show
increasing centralisation as the slope of the density decline
curve increases. We would broadly expect that city evolution
has gone in the reverse direction. An interesting feature of
these plots is that the high central devleopment is split into
two, an effect presumably reinforced by the population "hollow"
in the centre.

       We can now transfer this analysis to Leeds 30 x 900
system, although we present only one set of results (see Clarke
(1984) for a much more extensive set). In Figure 17 we examine
the effect of β variation (from 0.5 to 0.1). Note that the
pattern changes from one of highly dispersed centres, through
to a central dominated pattern, exactly as we would expect.

       So far the analysis has been concerned with cross-section-
al allocations of all retail capacity. We now proceed to
examine what happens when we use the model in an incremental
fashion. Figure 18 contains four of these increments based on
the previous pattern but with extra facilities (derived through
increasing demand) allocated. Note that we assume α increases
over time, β declines, and the population grows and spreads
out. The pattern changes from a rather unusual one with several
concentric rings to one with fewer centres, though with a
suburban orientation.

       The results presented in this section represent just a
small proportion of the experiments we have undertaken (and
once again we refer the reader to Clarke and Wilson, 1984),
where a full set is presented and where we pay particular
attention to model disaggregation.

5. THE EVOLUTION OF URBAN STRUCTURE: DISEQUILIBRIUM ANALYSIS

       We have already noted in section 3 that if revenue exceeds
costs in a zone that we would expect that centre to expand its
provision of retail facilities and vice versa. In that section
we were concerned in locating the equilibrium conditions when
revenue and costs balanced. In this section, we relax that
assumption and consider the evolution of spatial structure
through the use of differential and difference equations, and
using some results presented by May (1978) focus on the identification of the bifurcation points of these models. The work
reported here is a summary of that found in Beaumont, Clarke
and Wilson (1981b), Clarke (1984) and Wilson (1981).

       To recap, we assume that the dynamical behaviour of the
structural variables is based on profitability, that is

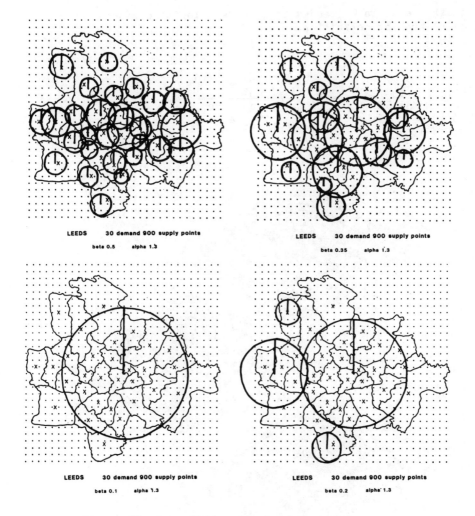

Figure 17. β variation for Leeds system

$$\dot{W}_j = \varepsilon(D_j - kW_j) \qquad (5.1)$$

where ε represents the rate of response to change in size to profit or loss. More specifically, we are concerned with the following difference equation

$$W_{jt+1} - W_{jt} = \varepsilon(D_{jt} - kW_{jt})W_{jt} \qquad (5.2)$$

(with the extra W_{jt} added to give us a logistic growth equation) and, by definition,

$$D_{jt} = \Sigma_i \, S_{ijt} = \Sigma_i \, \frac{e_i \, P_i \, W_{jt}^{\alpha} \, e^{-\beta c_{ij}}}{\Sigma_k \, W_{kt}^{\alpha} \, e^{-\beta c_{ik}}} \qquad (5.3)$$

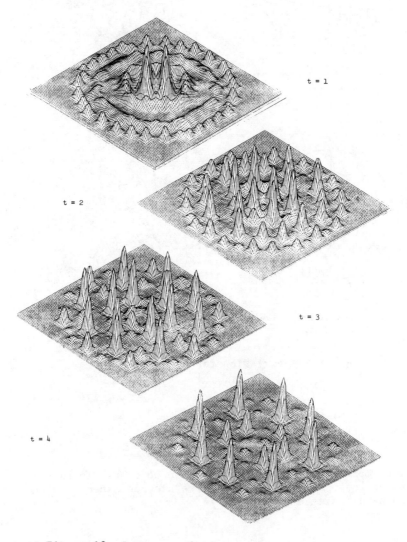

Figure 18. System evolution

Equation (5.2) can be rewritten as

$$W_{jt+1} = [(1+\varepsilon D_{jt}) - \varepsilon k W_{jt}]W_{jt} \tag{5.4}$$

Following May (1976) we know for a stable steady state to exist the following condition must be satisfied.

$$1 \leqslant 1 + \varepsilon D_{jt} \leqslant 3 \tag{5.5}$$

which is

$$0 \leqslant \varepsilon D_{jt} \leqslant 2 \tag{5.6}$$

Clearly εD_{jt} is non-negative, but is not necessarily less than 2. May then examined what occurred if this value exceeded 2; for values between 2 and 2.8495 periodic oscillations set in, between 2.8495 and 3.0 the oscillations become aperiodic and are termed chaotic; a value of εD_{jt} greater than 3.0 gives rise to divergent behaviour, leading to ultimate extinction. Clearly, each of these values represents a bifurcation point at which the nature of the solution (5.2) changes. However, it should be noted that D_{jt} is not a constant; it changes over time and is dependent on the spatial competition between centres. In this sense, the positions of the bifurcation points change in a complicated way.

This model is based on some fairly simple assumption: disaggregation between good types is needed and more realistic assumptions about producer behaviour, to incorporate effects such as time lags, are required. Some of these developments are discussed in Clarke (1984).

One of the interesting conclusions from the work of Allen (1981) is the importance of random fluctuations in the evolution of urban structure. To incorporate these effects into our model we simply add in a random term to equation (5.2), which becomes

$$W_{jt+1} - W_{jt} = \varepsilon[(D_{jt}+\upsilon) - kW_{jt}]W_{jt} \tag{5.7}$$

where υ is a random number distributed between + and − some value. We report on the effects this has on urban evolution in the next section.

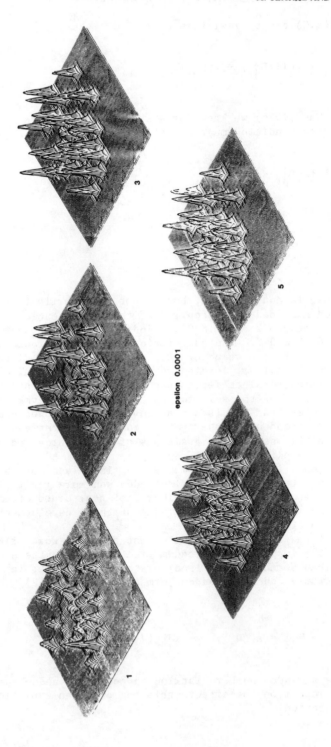

epsilon 0.0001

Figure 19(a)

epsilon 0.00020

Figure 19(b)

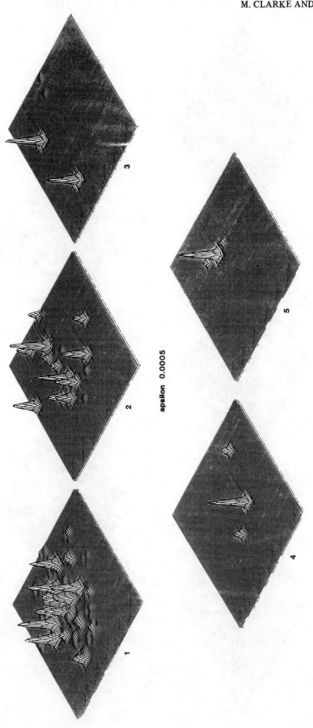

epsilon 0.0005

Figure 19(c)

6. NUMERICAL SIMULATIONS: THE DISEQUILIBRIUM MODEL

All the results presented in the section were derived using the Leeds 30 x 900 system. Firstly, we examine the pattern of development for different values of the ε parameter. Figure 19(a,b and c) represent the system evolution (starting from a uniform distribution of W_j) for $\varepsilon=0.0001$, $\varepsilon=0.0002$ and $\varepsilon=0.0005$. In the first case, the system appears to be heading towards an equilibrium solution and in the second and third case, other types of behaviour are demonstrated. This we can assume is due to the fact that the εD_{jt} term is passing through the bifurcation points outlined in the previous section. To confirm this we can plot the value of εD_{jt} against t (and also W_j against t) and examine if the value of this term passes through these points. This is done in Figure 20, for two zones (251 and 257). The three straight lines on the upper plots of εD_j-t are the three bifurcation points, 2, 2.8495 and 3. Four different ε values are examined, 0.0001, 0.00019, 0.0002 and 0.0003. For the first two values, the value of εD_{jt} does not cross through any of the bifurcation points and the W_j trajectory is smooth. However, when $\varepsilon=0.0002$, this is the case and oscillatory behaviour sets in. This becomes ever more pronounced when $\varepsilon=0.0003$. Thus we have identified an important system bifurcation when the value of ε passes between 0.00019 and 0.0002.

Finally, in this section, we present an example of the addition of fluctuations to our analysis. Figure 21 shows the pattern of facility location of 25 iterations of the model run under identical parameter values with a random term, υ, distributed between +10 and -10 (a relatively small value in comparison to D_{jt}). Each run used a different random number stream and the effects on the patterns is quite noticeable. We must therefore concur with Allen that fluctuations may have an important role to play in the evolution.

7. CONCLUSIONS

We hope we have effectively demonstrated that the analysis of bifurcation phenomena provides interesting new insights into the evolution of urban spatial structure. Significant progress has, we believe, been made, but the research agenda is still of considerable length. For example, we need to examine the residential and industrial location problems in an analogous way.

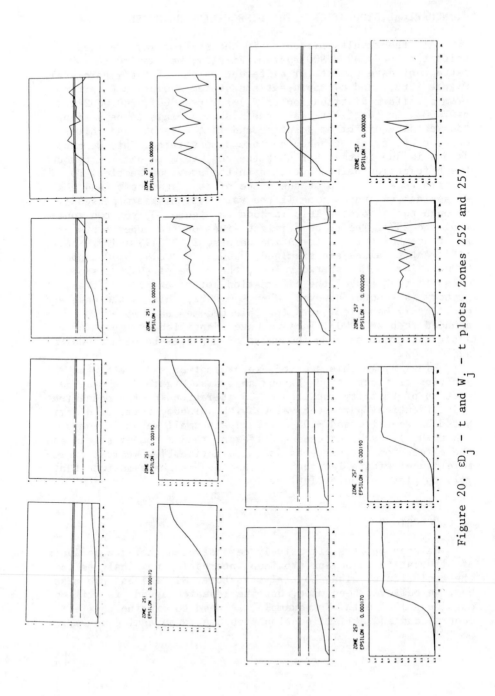

Figure 20. $\epsilon D_j - t$ and $W_j - t$ plots. Zones 252 and 257

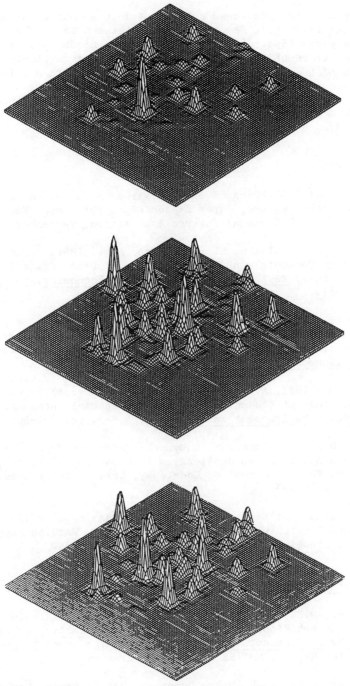

Figure 21. System evolution after 25 iterations:
introduction of fluctuations

Then we need to link these subsystems into a comprehensive modelling framework. The relationship between the methods used in this paper (Spatial Interaction models) needs to be related to other types of approaches, perhaps most interestingly the methods of the "New Urban Economists" (see Richardson, 1977). The there remains the difficult task of investigating how these models should be used in a planning context. This will probably require new forms of data, but it is, of course, the natural and desired outcome of this work.

REFERENCES

Allen, P.M.: 1981, Urban Evolution viewed as a self-organising non-linear system, Paper presented at the Annual Meeting of the British Regional Science Association, September 1981.

Beaumont, J.R., Clarke, M. and Wilson, A.G.: 1981a, Changing Energy Parameters and the Evolution of Urban Spatial Structure, Regional Science and Urban Economics, 11, pp. 287-315.

Beaumont, J.R., Clarke, M. and Wilson, A.G.: 1981b, The dynamics of urban spatial structure: some exploratory results using difference equations and bifurcation theory, Environment and Planning, A, 13, pp. 1473-1483.

Clarke, M.: 1981, A note on the stability of equilibrium solutions of production constrained spatial interaction models, Environment and Planning, A, 13, pp. 601-605.

Clarke, M.: 1984, Integrating Dynamic Models of Urban Structure and Activities: an application to urban retail system, Ph.D. Thesis, School of Geography, University of Leeds (forthcoming).

Clarke, M. and Wilson. A.G.: 1981, The dynamics of urban spatial structure: progress and problems, Working Paper 313, School of Geography, University of Leeds.

Clarke, M. and Wilson, A.G.: 1984, The Dynamics of Urban Spatial Structure, Croom Helm, London (forthcoming).

Dendrinos, D.S. and Zahavi, Y.: 1980, Critical Points in Mode Choice Behaviour, in: Dendrinos, D.S., Catastrophe Theory in Urban Transport Analysis, U.S. Department of Transportation Report, DOT/RSPA/DPB-25/80/2.

Forrester, J.W.: 1969, Urban Dynamics, M.I.T. Press, Cambridge, Mass.

Harris, B. and Wilson, A.G.: 1978, Equilibrium values and dynamics of attractiveness terms in production constrained spatial interaction models, Environment and Planning, A, 10, pp. 371-388.

Harris, B., Choukroun, J-M. and Wilson, A.G.: 1981, Economies of scale and the existence of supply-side equilibria in a production-constrained spatial interaction model, Working Paper 305, School of Geography, University of Leeds.

Huff, D.L.: 1964, Defining and Estimating a trading area, Journal of Marketing, 28, pp. 34-38.

Isard, W. and Liossatos, P.: 1977, Models of Transition Processes, Papers, Regional Science Association, 30, pp. 1-14.

Lakshmanan, T.R. and Hansen, W.G.: 1965, A retail market potential model, Journal of the American Institute of Planners, 31, pp. 134-143.

May: 1976, Simple mathematical models with very complicated dynamics, Nature, 261, pp. 459-467.

May: 1978, The Evolution of Ecological Systems, Scientific American, 239, No. 3, pp. 161-175.

Ravenstein, E.G.: 1889, The Laws of Migration, Journal of the Royal Statistical Society 52, pp. 241-301.

Richardson, H.W.: 1977, The New Urban Economics: and Alternatives, Pion, London.

Wagstaff, J.M.: 1978, A possible interpretation of settlement pattern evolution in terms of catastrophe theory, Transactions, Institute of British Geographers, New Series, 3, pp. 165-178.

Wilson, A.G.: 1981, Catastrophe Theory and Bifurcation: Applications to urban and regional systems, Croom Helm, London.

Wilson, A.G. and Clarke, M.: 1979, Some illustrations of catastrophe theory applied to urban retailing structure, in: M. Breheny (ed.), Developments in Urban and Regional Analysis, Pion, London.

Zahler, R.S. and Sussman, H.J.: 1977, Claims and accomplishments of applied catastrophe theory, Nature, 269, pp. 759-763.

BIFURCATION SETS - AN APPLICATION TO URBAN ECONOMICS

M. Prévot

Institute of Mathematical Economics
University of Dijon

1. INTRODUCTION

Mathematical models capable of representing drastic transformations of form and structure - the transformations to which
the human spirit is most sensitive - have so far been little
explored. Until lately, shock therapy, applied in mechanics,
was the only theory that could take account of such phenomena;
however, the mathematician R. Thom has recently opened up a
wide field of research by developping a general theory of
morphogenetic models.

Thom's idea is to construct mathematical models representing the physical, biological or linguistic world of shapes in
their continuous movement of birth, growth and decay. "Le
propre de toute forme de toute morphogénèse est de s'exprimer
par une discontinuité du milieu" (Thom, 1972).

So far, this new theory has found little application
either in economic science in general or in spatial economics
in particular; for that reason the present paper sets out to
introduce in an informal way the concept of bifurcation sets
and to give an example of its application to urban economics.

2. FORMAL CONCEPTS

Most theories in spatial economics assume a metric space;
we shall however define a space which is just a little more
general: a differentiable manifold; such a space being locally
equivalent to an euclidian space. Then we will define the
concept of germs and stability. The absence of stability leads

101

M. Hazewinkel et al. (eds.), Bifurcation Analysis, 101–114.
© *1985 by D. Reidel Publishing Company.*

us to define the universal unfolding of a function and the
concept of bifurcation sets.

2.1. Differentiable manifolds

2.1.1. Smooth application. A map f of an open subset U of
R^n into R^m is smooth if and only if it is infinitely differentiable. More generally if $X \subseteq R^n$ is a subset f: $X \subseteq R^n$ is said
to be C^∞ if there exists a C^∞ extension $\tilde{f}$: $U \to R^n$ for some open
subset U of R^n containing X. We denote with $C^\infty(n,m)$ the
function space of the smooth applications of R^n to R^m.

There is an obvious analogous notion to that of homeo-
morphism in topology.

2.1.2. Diffeomorphism. A smooth map f of X into Y, respectively subsets of two euclidian spaces, is a diffeomorphism if
and only if f is bijective, f^{-1} is smooth; then X and Y are
called diffeomorphism: i.e.

$$X \text{ and } Y \text{ diffeomorphic} \Longleftrightarrow f \varepsilon C^\infty(X,Y) \text{ f bijective } f^{-1} \varepsilon C^\infty(X,Y)$$

$$(2.1)$$

The concept of diffeomorphism is by definition a global
property, for it applies to the whole space; by taking only a
local property it is possible to define differentiable manifolds.

2.1.3. Differentiable manifolds. Let $X \subseteq R^n$, X is a
differentiable manifold if and only if X is locally diffeo-
morphic to R^p, p is the dimension of the manifold. In other
words, there exists an open covering of X such that $f(U_i)$
defines a diffeomorphism of U_i on to a subset of R^p. Thus by
definition a differentiable manifold is obtained from local
properties.

Obviously, differentiable manifolds can be defined on
spaces more general than euclidian ones; nevertheless we shall
base ourselves on that definition, which is the only one used
in spatial economics.

Let X be a differentiable manifold, a point $a \varepsilon X$ and let us
take a neighbourhood U of a; some among all numerical functions
defined on X will take the same value on U. Consequently, instead of considering the whole set of differentiable functions
of X to R of which one wants to study the properties in a
neighbourhood U of a, one can just study those that have
different values on U; so, these functions can be put into
equivalent classes, which takes us to the concept of germs.

2.2. Germs: definition

Given two differentiable manifolds with respective dimen-

sions p and q and $L(X,Y)$ the function space of continuous
applications of X to Y defined in the neighbourhood of a given
point in X, two functions f and g, elements of $L(X,Y)$, are
equivalent if and only if they take the same values in some
neighbourhood U of that point. The relation thus defined is an
equivalence relation; an equivalence class of which f is representative is the germ of function f; $\varepsilon(p,q)$ denotes the space
of all germs at a given point.

Note that the space of germs $\varepsilon(p,q)$ provided with the
operations "addition" and "multiplication", has an unitary
communative ring structure.

2.3. Universal unfolding

2.3.1. The problem. Assume that the manifold X describes
the evolution of parametrics x, which characterise a certain
process. To every x is associated a numerical function $f(x)$
measuring the energy, the entropy or the potential of the
system in state x; this function can be, for example, the
utility of the consumer or the profit of the producer.

If the process starts at $a\varepsilon X$, $f(a)$ = Cste because response
always needs some time, we suppose that $\partial f(a)/\partial x_i = 0$ and that
a is a critical point; the a priori assumption that this point
is non-degenerate and isolated, i.e. that there are no other
critical points in a neighbourhood of a, follows then
logically.

No matter with how much care the study of a phenomenon is
undertaken the initial data will vary slightly; that small
variation in the data is expressed in the existence of a
diffeomorphism h, close to the identical application which
transforms x into $h(x)$. Thus, the results are prone to errors.
So to pass results from one experience to another, still
another diffeomorphism has to be introduced, namely h' from R
to R also close to the identical application.

By definition on application $f: X \to R$ is stable if and
only if, for all f' sufficiently close to f, there exist two
diffeomorphisms h and h' close to the identity applications, so
that the following diagram commutes

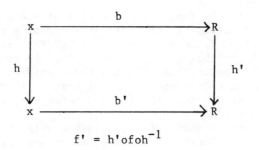

$$f' = h'ofoh^{-1}$$

Figure 1.

This definition does no more than present in a formalised way the intuitive concept of stability: every map f' close enough to it is the same as f up to "recoordinisations" of source and target space; mark, by the way, that this property is a global one.

Remark

Obviously, the defintion can be extended to the case where the target set is a differentiable manifold Y.

When a map f: X → Y is not stable, one considers "unfoldings" of f, that is families of mappings $F: XxR^r → YxR^r$ which are level preserving, i.e. $F(Xx\{u\} \subset Yx\{u\})$ and such that $F(x,o) = (f(x),0)$, F is a family of mappings $F_u: X → Y$ $(F_u(x),u) = F(x,u)$ projected onto Y parametrised by $u \varepsilon R^r$ such that F_o = f. Such an unfolding is <u>versal</u> if any other can be obtained from it (by parametrisation) in a way to be specified below and it is <u>universal</u> if it is versal and the parameter dimension r is minimal. All these notions also make good sense for map-germs. Roughly a versal unfolding of a map-germ contains representatives (parametrised in a smooth way) of all map-germs f': X → Y close to f.

This also permits a neat way to give the right definition of stability of a map-germ: the map-germ f: X → Y is stable if f: $Xx\{0\}$ → $Yx\{0\}$ is a versal unfolding. It is to be noted that this differs from the obvious "germ-version" of the global definition of stability given above, which as it turns out does not give the right definition of stability of germs. (Then all germs defined by a stable map would not be stable; also there are difficulties with respect to the topology on the space of map-germs).

It remains to give the technical definition of how one unfolding is obtained from another by reparametrisation or, as one usually says, is induced from another. This can probably be skipped without essential harm to one's understanding of what follows. Let XxR^r → YxR^r be an unfolding of f: X → Y and let G: XxR^s → YxR^s be another unfolding of f. Then G is induced from F if and only if there exists a smooth map e: R^s R^r,

$v \to u=e(v)$ and families of diffeomorphism parametrised by
$v \epsilon R^s$, $\phi_v: X \to Y$, $t_v: Y \to Y$ depending smoothly on v such that

$t_v G_v(u) = F_{e(v)}(\phi_v(x))$. For germs of unfoldings represented by
F, G everything remains the same so that $v \to e(v)$, $v \to \phi_v$, $v \to t_v$

need only be defined in a small neighbourhood of $0 \epsilon R^s$, ϕ_v, t_v
need only be local diffeomorphisms (germs of diffeomorphism)
and the equality is to be interpreted as an equality of germs.

This very recent theory has close links with the theory of
rings, the ideal of a ring and with the preparation theorem of
Malgrange, which is an extension of the Weierstrass's theorem
on the approximation of a function by polynominal; it can be
shown that nearly every application possesses a versal
unfolding.

3. BIFURCATION SETS

3.1. Model

The elements studied are located in a space W, which is a
differentiable manifold; the local state of the process in point
$w \epsilon W$ is defined by a number of parameters $\underline{x} = \{x_1, \ldots, x_n\}$.
The set of parameters in w, which can characterise the
local state of the process, is represented by a point P of the
parameter space X; X is called the manifold of internal states.
The process may be supposed homogeneous enough for the manifold
of internal states X to be independent from the point w chosen;
X is often supposed to be a compact differentiable manifold of
dimension n.
When the point w moves at a regular pace towards point w',
point P moves towards P' with usually a similar regularity.
There are points of w, however, where that regularity disappears; we shall indicate by K_w the set of discontinuities of
the morphogenetic process we are considering. The problem now
is to recognise the shape of that set of discontinuities K_w as
well as its topological nature.
Let us take a closer look at what happens if P moves
towards P' in a regular way. Point P" characterising the local
state of the process between P and P' lives on a curve E, the
path along which the process evolves, linking P to P'. If w' is
very near to w in the set w, E can be projected onto X_w; P is
projected on P which describes a path in X_w. The assumption is
that this path is independent of the route joining w to w' in
w. The movement of the point can be described either numerically by a differential equation or geometrically by the vectors touching the path in each point, these vectors are indeed
those of the moving speeds. In this differential system a vector field is defined by the local regime of the process. Very

generally we assume that this regime is defined by a vector
field χ on manifold X.

As occurs regularly in mechanics, a potential function
F(x) may be associated with this system, giving its energy in
each point x. By definition the field   then has the compo-
nents $-\partial F/\partial x_i$. The system being given by a system of differential equations or vector fields, F can be said to be parametrised by w and written in condensed form F_w. Under those
conditions the stable regime characterised by P is defined by
the minima of F_w. In the general case F admits several minima
so that the choice of the stable regime is not uniquely defined
and can only be effectuated by a series of conventions.

Finally, the evolution of F_w as a function of w remains to
be specified. If the potential initially shows a singularity,
an attempt should be made to extend the potential such as to
preserve the qualities of homogeneity, uniformity and stability
as point w moves within the set W. According to Guckenheimer's
theorem there is a family of functions possessing these qualities, namely the unversal unfolding $F(x,u_i)$; the problem is
what realisation of $F(x,u_i)$ to choose to represent the potential in a single point w. Let U be the space to which the
parameters u_i belong and denote by G the mapping defined by

$$W \rightarrow U$$

$$w \rightarrow u_i \epsilon(Gw)$$

(3.1)

G being the growth or the evolution wave of the morphogenesis.
The corresponding field $-\partial F/\partial x_i(G(w))$ is the metabolic field of
the phenomenon.

If G is continuous, as it may justly be supposed to be,
the universal unfolding, of which the number of parameters u_i
is less than four, are given by the elementary potentials of
Thom.

The name "catastrophy set" is given to the set $K_u = K_i$,
the value of the parameters for which the number of minima
changes; to the extent that U and v have been identified,
$K_u = K_w = K$.

3.2. Bifurcation set

3.2.1. Definitions. An "algebraic set" in R^n is understood
to be a set of points $\underline{x} = \{x_1, \ldots, x_n\}$ of R^n, solutions of m
polynominal equations $\bar{P}_i(x) = 0$.

Such a set is the finite union of disjoint sets that are
differentiable manifold called strata. Strata can be classified
in a general way by their dimensions. Whitney has developed
conditions of linkages among strata.

Essentially, set B of parameter values for which the
number of minima of the potential changes is called the bifur-

cation set; B is the set of unstable phenomena and contains the
set K of catastrophies. So what is the difference with the
catastrophy set?

If the singularity is of corank2, B is defined by

$$\frac{\partial F}{\partial x} = \frac{\partial F}{\partial y} = 0 \tag{3.2}$$

The corresponding point being singular,

$$\frac{\partial^2 F}{\partial x^2} \frac{\partial^2 F}{\partial y^2} - \left(\frac{\partial^2 F}{\partial x \partial y}\right)^2 = 0 \tag{3.3}$$

the singularity being degenerate.

It is easy to see that slight modifications of the
structure completely transform the aspect of the phenomenon.

By the conflict set is understood the set C of values of
parameters for which two minima have the same value. The two
sets are stratified.

The two definitions have to be translated in terms of
functional spaces; the idea of jets introduced by Ehresmann
will enable that concept to be clarified.

3.2.2. Jet space.
<u>Definition</u>

Let X and Y be two differentiable manifolds, $x \epsilon X$, f and g
two smooth applications from X to Y such that $f(x) = g(x) = y$.

f has a first order contact with g if and only if
$(df)x = (dg)x$, $(df)x$ indicating the matrix of partial derivatives at point x.

f has a contact of order k with g in point x, if and only
if $(df)x$ has a contact of order k-1 with $(dg)x$; the relation
indicated as "having a contact of order k" is written $f \underset{k}{\sim} g$. This
relation is one of equivalence.

$J^k(X,Y)(x,y)$ is the quotient $C^\infty(X,Y)/\underset{\sim}{} $ at point $(x,y=f(x))$.

When the whole space is considered, one defines the k-jet
as follows: $J^k(X,Y) = \underset{x \epsilon X}{} J^k(X,Y)_{(x,f(x))}$.

In that case functions f and g have the same Taylor
expansion up to order k; let us denote by $J^k$of the jet of order
k defined at the origin, of which f is representative.

$$J^k of = f(o) + (x-o)f'(o) + \ldots + \frac{(x-o)^k}{k!} f^k(o) \tag{3.4}$$

$$J^k of = J^k og \quad \text{if g belong to this k-jet.} \tag{3.5}$$

Let us indicate by $J^k(n,m)$ the set of k-jets from R^n to R^m taking origin to origin. The jet spaces thus defined have the structure of a vector space. $J^1(n,1)$ has as basic vectors x_i, i=1,...,n.

$J^2(n,1)$ has as basic vectors x_i; $x_i x_j$ and so forth. The vector spaces can be canonically projected one to each other, J^{k+1} onto J^k and so on. $J^k = p^k(J^{k+1})$ where p^k represents the projection mapping.

These spaces can be provided with a topology i. Let us denote by sup $(J^k f)$ the maximum as x runs through a neighbourhood U of the origin of $|f(x)| + |f'(x)| + ... + |f^k(x)|$. An open neighbourhood U_e of f can be defined by $g \varepsilon U_e$ if and only if $|$ sup $J^k f -$ sup $J^k g | < e$; the topology thus defined by its neighbourhoods is called the C^k topology.

Preferable to a direct description of $C^\infty(X,Y)$ is a step by step procedure, by which the succession of local jet spaces in the various points x of X are described.

In the first order, that is to say in $J^1(n,1)$ functions f are easy to classify topologically. One of the coefficients of f'(o) is not zero at the origin, which implies that at least one of the partial derivatives is not zero; by the classical theorem of implicit functions, function f is locally transformable into a linear function by changing curvilinear coordinates. In other words, whatever the values of coefficients for order higher than one, the form of the function is univocally fixed.

By contrast, if all coefficients are zero, the topological nature of function f is more difficult to specify; in that case the conditions $\partial f_j / \partial x_i = 0$ represent in J^1 the bifurcation set K^1.

In the second order, the points of $J^2(n,1)$ projected by p^1 outside K^1 have their destiny i.e. their topological type perfectly fixed.

By contrast, the points of J^2 projected to a point of $K^1 \subset J^1$ admit of the following classification.

Consider the quadratic form associated with the matrix

$$\left(\partial^2 f_k \right) / \left(\partial x_i \partial x_j \right)$$ when that quadratic form is negative or posi-

tive definite by the Morse theorem the addition of terms of an order higher than two does not change the topological type of the function, which remains that of a quadratic form.

Things are different, however, if that quadratic form is degenerate; for such quadratic forms the topological type of the function is no longer fixed and these points in $J^2 \cap p^{-1}(K^1)$ form the bifurcation set K^2.

It is possible to continue the study and define for each order r an algebraic subset K^r.

Theorem

The space of jet $J^r(n,m)$ contains a bifurcation set K^r.

Since K^r is an algebraic and stratified set, the stratification can be described as follows: consider the series $J^1 \supset J^2 \ldots \supset J^r \ldots$; the application p^k induces among the bifurcation sets the inclusion relation $K^1 \supset K^2 \ldots \supset K^r$; $p^i(K^i) \quad K^{i-1}$

Example

$$J^1 of = a_1 x_1 + a_2 x_2 \qquad a_i = \partial f/\partial x_i(o) \qquad (3.6(a))$$

$$K^1 = \{a^1 \epsilon X, \ a^2 \leftarrow X \qquad a_1 = a_2 = 0\} \qquad (3.6(b))$$

$$J^2 o(f) = a_1 x_1 + a_2 x_2 + b_1 x_1^2 + 2b_2 x_1 x_2 + b_3 x_2^2 \qquad (3.7(a))$$

$$K^2 = \{b_1, b_2, b_3; \ b_2 - b_1 b_3 = 0\} \bigcap K^1 \qquad (3.7(b))$$

Note that according to the preceding property it is a matter of a series of embedded spaces each of which is projected on the previous one; such a system is called a projective limit in mathematics. In certain cases such embedded spaces can be reduced to a space of fixed given dimension by considering the singularitites of fixed dimension and their universal unfolding.

As has been shown above the applications f and g on $J^r(n,m) - K^r$ are of the same topological type. Sometimes the jet may be structurally stable; so that any sufficient small deformation of a mapping f is topologically equivalent to f. For example, a critical quadratic non-degenerate point is structurally stable in the neighbourhood of that point; if that is not so, two cases may present themselves.

(1) Arbitrarily small perturbations may present an infinite number of different topological types; the singularity represented by f is said to be of infinite codimension; no finite order jet of f can be determinate.

(2) The deformations of f present only a finite number of topological types; the singularity is said to be of finite codimension q. A sufficiently high order jet of f then is determinate and its codimension is q; in that case the mapping f can be immersed in a family F of deformations with q parameters such that the mapping F defined on the set $R^n x R^q$ is structurally stable. In particular, a structurally stable jet is of codimension zero; the unfolding is the organising centre f itself.

Example
$$f(x) = x^4/4 \tag{3.8}$$

$$F(x,u,v) = \frac{x^4}{4} + u\frac{x^2}{2} + vx \tag{3.9}$$

$$J^1 oF = vx \tag{3.10(a)}$$

$$J2oF = ux + u\frac{x^2}{2} \tag{3.10(b)}$$

$$J^3 oF = vx + u\frac{x^2}{2} \tag{3.10(c)}$$

Thus, this example shows immediately that the datum of jet $J^2 oF$ completely determines the unfolding of $f(x) = x^4/4$.

3.3. Conclusions

Apparantly from these considerations, two cases may present themselves.

On the one hand, there may exist a jet of f which is determinate in the function space of mappings $C^k(n,m)$. There also exists a bifurcation set Y with a stratified structure. A singularity of f at x of dimension q is a stratum of Y of co-dimension q in the space $L(n,m)$ of the mappings from R^n to R^m. The universal unfolding is nothing but a fragment of q, trans-versal at point x to the stratum of f. Apparently, then, the universal family is not unique, but only defined up to a stratified isomorphism.

On the other hand, there may be no finite jet of f that is determinate; so let Σ be the set of jets with non-finite co-dimension corresponding to applications without universal unfolding and not structurally stable. $J^r(n,m)$ can be strati-fied according to the value of codimension k attached to any mapping f of jet J^r of $\epsilon\ J^r(n,m) - \Sigma$.

Imagine that to each stratum J^r there is associated a stone of thickness r and of uniform colour. Those stones are put one on top of the other by increasing order of thickness. Each stone is layered by pictures of stratified bifurcation and catastrophy sets for each type of universal unfolding, joined together while fulfilling the Withney conditions.

4. ECONOMIC APPLICATIONS

4.1. Binary choice

Consider in an urban space a binary choice between two transportation modes determined by an utility function $U(x,u,v)$, where x is the choice variable, u and v being two exogenous parameters of the model.

$x < 0$ indicates the choices of transport mode 1, while $x > 0$ indicates mode 2. Now before introducing a potential, we have to determine the minima of a function; that is why we define $F = -U$, presenting itself as a disutility, and construct a model in which the choices are made such as to minimise the function, the way it is done in potential theory.

Leaving several possibilities to the consumer, we have to introduce the universal unfolding of a function that has a singularity at the origin. In our simple case we shall take the most simple function that presents a degenerate singularity at the origin, namely

$$F(x,u,v) = \frac{x^4}{4} + \frac{ux^2}{2} + vx \qquad (4.1)$$

The stationary values of F are given by

$$\frac{\partial F}{\partial x} = x^3 + ux + v = 0 \qquad (4.2)$$

The bifurcation set is then given as in the previous example by the zeroes of the second derivative

$$\frac{\partial^2 F}{\partial x^2} = 3x^2 + u = 0 \qquad (4.3)$$

Elementary calculus supplies the bifurcation set

$$4u^3 + 27v^2 = 0 \qquad (4.4)$$

To specify the nature of our model, let us introduce differential cost; for a specific trip made by an individual, let C^1 and C^2 be the transportation costs associated with modes 1 and 2, and $\Delta C = C^2 - C^1$ represent the difference in cost. Variables u and v being exogenous to the model, we may take

$$u = f(\Delta C) \qquad\qquad v = g(\Delta C) \qquad\qquad (4.5)$$

By the assumptions made, we can construct functions f and
g with the following properties:
1) With ΔC large and positive, mode 1 is chosen and x is
   negative.
2) With ΔC large in absolute value but negative, mode 2 will
   be taken and x is positive.
3) If $0 < \Delta C < k$, mode 1 is the more readily chosen but mode 2
   is also possible.
4) If $-k < \Delta C < 0$, model 2 will be the one more readily chosen
   but both choices are possible.
   Given the symmetries in the model one necessarily has to
take an even function for f and an odd one for g.
   Our study being a local one, an approximation can be given
of the functions f and g with the help of a Taylor formula; in
this way one obtaines

$$u = a(\Delta C)^2 - b \qquad\qquad (4.6(a))$$

$$v = d(\Delta C) \qquad\qquad (4.6(b))$$

where a, b and d are positive constants. In view of the symmetry of the function u and v, it is easy to verify that the
five conditions mentioned earlier are satisfied; the critical
values are indeed obtained by solving the equation

$$4(a(\Delta C)^2 - b)^3 + 27d^2(\Delta C)^2 = 0 \qquad\qquad (4.7)$$

This equation is a sixth degree one that can be reduced to
a third degree one by putting $(\Delta C)^2 = \lambda$

$$4(a\lambda - b)^3 + 27d^2\lambda = 0 \qquad\qquad (4.8(a))$$

$$4a^3\lambda^3 - 12a^2b\lambda^2 + (12ab^2 + 27d^2)\lambda - 4b^3 = 0 \qquad\qquad (4.8(b))$$

It is easy to verify that:
$$P(\lambda) = 4a^3\lambda^3 - 12a^2\lambda^2 + (12ab^2 + 27d^2)\lambda - 4b^3 \text{ is always}$$

increasing, so $(p(\lambda)$ admits a single real root $\tilde{\lambda}$.

Because $p(o) = -4b^3 < 0$.$\tilde{\lambda}^{\cdot} > 0$. The condition $\tilde{\lambda} > 0$ leads
to $b > 0$ as we have seen above.

4.2. Generalisations

The difficulty which the model presents is the estimation of constants a, b and d. They can be estimated by econometric methods or in the manner presented in Fustier's Ph.D. thesis, by drawing homogeneous subgroups from the population set.

5. CONCLUSIONS

Bifurcation set theory has thus found justification in spatial economics. In spatial economics use is often made of elementary potential or of the differentiable system representing the evolution of the phenomenon. Moreover, certain properties of universal unfolding can be used. For instance, universal unfoldings with five or more parameters form in space families of differentiable functions with five parameters, which are an open dense subset of Whitney's C^∞ topology. In that case, Morse's theory allows an approximate and qualitative vision of the phenomenon.

In many examples the cusp appears as universal unfolding, which is acceptable as a simplification, as we have seen in our earlier example.

REFERENCES

Abraham, R.: 1972, Introduction to morphology, Quatrième rencontre entre mathématiciens et physiciens, Département de Mathématique de Lyon, Vol. 4, pp. 38-114.

Bruter, C.P.: 1973, Sur la nature des mathématiques, Gauthier-Villars, Paris.

Fustier, B.: 1981, Les interactions spatiales en économie, Collection de l'I.M.E., Vol. 21.

Guillemin, V. and Pollack, A.: 1974, Differential topology, Prentice Hall, New Jersey.

Guckenheimer: 1973, Bifurcation and catastrophe, in: Peixoto (ed.), Dynamical Systems, Academic Press, New York, 1973.

Takens, F.: 1972, Singularities of functions and vector fields, Nieuw Archief voor Wiskunde (3), XX, pp. 107-130.

Thom, R.: 1970-1971, Modèles mathématiques de la morphogénèse, Séminaire IHES 1970-1971.

Thom, R.: 1972, Stabilité structurelle et morphogénèse, Ediscience, Paris.

Thom, R.: 1973, Théorie des catastrophes, état présent et perspectives, Manifold Spring, Mathematical Institute, Warwick University.

Young Chen Lu: 1976, Singularity theory and an introduction to catastrophy theory, Springer Verlag, New York.

NERVES AND SWITCHES IN CONFLICT CONTROL SYSTEMS[1]

Doede Nauta jr.

Department of Philosophy and History
Polytechnic University Twente

1. INTRODUCTION

Bifurcation has a very wide range of application.
Poseidon's trident and the mathematician's delight, n-forks,
surely are included in the theme of this book, provided only
that the dents of the n-furcation, c.q. trifurcation, are dynamically interrelated – something which is not the case with the
usual physical manifestation of Poseidon's fork. Any n-furcation, as you know, can be reconstructed as a combination of
bifurcations which partly overlap. Thus in section 4 we will
have to deal with trifurcations like:

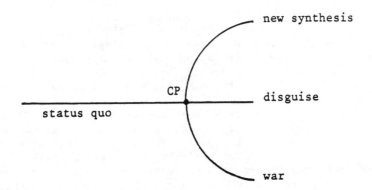

Figure 1. Basic trifurcation. CP is short for crisis-
point or conflict-point

M. Hazewinkel et al. (eds.), Bifurcation Analysis, 115–140.
© *1985 by D. Reidel Publishing Company.*

This basic trifurcation can be reconstructed as a subbi-furcation's resultant (dotted lines in Figure 2) combined with the rest of the trifurcation (Figure 3).

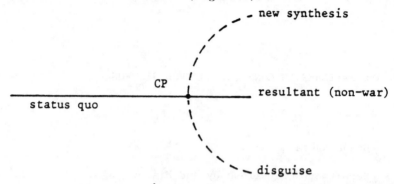

Figure 2. Artificial composition of the resultant of a bifurcation

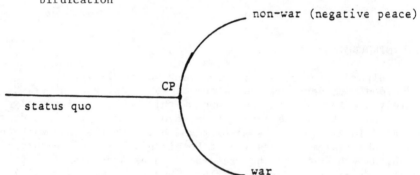

Figure 3. Bifurcational representation of the basic trifurcation

An important conclusion can be drawn from this simple pro-cedure: a dynamic process, such as perception or anything going on in society, complex as it may be, can always be adequately reconstructed as the interplay of just two (compound) polar forces. The well-known fact that identification presupposes differentiation, that is the difference of two things, is a case in point.

The fact, however, that there are three dents to a bifur-cation reminds us of the circumstance that all things consider-ed, taking into account the past, present and future of a pro-cess, you need three things. This has been systematically ela-borated in Charles Peirce's constructive pragmatism (CP); as a matter of fact, Peirce introduced a stylised version of the bifurcation in Figure 3, viz. $\prec$, as a new formal symbol in his (topo)logical system, where it has occupied a central place ever since.[2] According to Peirce all things in heaven and on

earth can be reconstructed in terms of three categories:
'firstness', 'secondness', and 'thirdness'. The reason why you
need at least three things ('categories') is demonstrated by
the difference between the loosely combined graph: $\rightleftharpoons$
which lacks singular points, and the integrated CP-graph: $\prec$
with its singular point.

Accordingly, Peirce's (topo)logical universe has three
elements: points ('firstness', pure qualities or possibilities), lines ('secondness', relations or existential connections) and 'spaces', i.e. decision-lines with nodal points
('thirdness', interrelations or habits).[3]

So much for the formal background of the contribution and
its connection with the general theme of the book; keywords in
the presentation which connect with 'bifurcation' are: crisis,
conflict and switch.

As to the content of the contribution, it focuses around
the concept of conflict; its motivational background may be
rendered as follows. The current equilibrium of nuclear deterr-
ence is an outstanding example, on the macro-level, of a conflict control system with its own 'nerves and switches'.

According to the official ideology of the NATO-officers,
ministers of defense, etc. this dynamic equilibrium is the outstanding preventor of war. It is said to have prevented, or at
least delayed, world war 3 for nearly 35 years. The critical
question which arises here is whether this conflict control
system does not in turn incorporate a menace greater than human
history has ever witnessed, all kinds of menaces of war
included. The conclusion is clear to everyone who is not a
party in the decision-making process of armament in the context
of cold war; thus, an exposition of this specific subject has
no sense.

Still there is much sense in the more general problem
whether it is possible to design conflict control systems that
are substantially less menacing themselves than the outbursts
that they try to prevent or to canalise; in order to come to
grips with this problem we need a conceptual framework for
discussion and analysis. In this connection we will have to
consider, for example, whether conflict is a 'positive' or
'negative' phenomenon, and what its possible outcomes are; also
who or what kind of institution will operate conflict control
systems, and many questions of many kinds more have to be studied in this context; here there is only room for a discussion
of some of these issues. On the whole, abstraction is made of
the many different contexts in which conflict control systems
apply. In order to introduce some of these contexts, we can
think of conflict control particularly in connection with what
is going on in Poland at the moment; and more generally we
should mention social turbulance, economic war, education,
psychiatry ('autoconflict'), crime and law, guerrilerros and
dictators, and also 'nature invented solutions' in the struggle

for life.

    'Jurisdiction' may be said to be one of the most experienced and elaborated conflict control systems in use.

    The content of this paper is connected with the general theme of this symposium, bifurcation theory, in the following way. Social units, such as states and institutions, are dynamic systems which can be pushed into disequilibrium by the effects of exogenous or endogenous conflict. What happens then is that one out of a finite set of possible semi-stable solutions of the system is selected. This selection is partly a matter of <u>conscious</u> design (decision making). This is what distinguishes my paper from most of the other contributions to bifurcation theory. The conceptual analysis that follows is meant to enable and to stimulate <u>goal-directed</u> selection within the given problem context. In view of this two main classes of goals are indicated here: (1) war-prevention or conflict control, (2) (stimulation of the) design of a new synthesis that fulfils the conditions of a semi-stable solution of the system and that functions as a first approximation of 'peace as a positive conception'.

2. NERVES AND SWITCHES

    The first common sense remark to make here is that in common parlance 'nerves and switches' has associations like 'nervousness' and 'nervous breakdown' which it is relevant to take advantage of when you are concerned with conflict control systems. It makes you entertain the idea of switching on an emergency-brake, it makes you consider the implementation of emergency signals, it makes you take seriously the reality of passions and emotions, and how to control them.

    At the beginning of a philosophical lecture some anthropological reflection is not out of place. In particular, in view of the complex subject matter that is before us, it is worth while to stop for a moment at the following meditative train of thought associated with 'nerves and switches'.

    To stay in the train for better and for worse is not man's fate: he can always switch on the emergency brake and leave the system. He can put an end to it, or design a new one. Man, by virtue of his excentricity, is the only creative creature, the only living being that designs - for instance - its own conflict control systems. As a result of this uniquely ambiguous position man is 'condemned' to freedom: he has the freedom to get out of his own systems, to leave any given system and to step out of it. Suicide and self-sacrifice (for instance in the form of a Gandhian hunger-strike) are extreme models of this fundamental capacity for lateral thinking and acting; at the same time they are examples for the ambiguous sign character of man's existence. <u>Systems should suit man, not man systems.</u>[4)]

From common parlance we now proceed to a more specific use
of the terms 'nerve' and 'switch'. We will use the term 'nerve'
in a narrow and in a broad sense. In the narrow sense 'nerve'
will be a substitute for the technical term 'axon' as it is
used in neurophysiology; axons represent the infrastructure for
signal-traffic in the nervous system. Next to axons you have
synapses, a kind of micro-switches or neuronal decision-nodes
(see Figure 4). The system of axons represents the fixed part
of the nervous system as information processor, the system of
synapses represents the dynamic part of it.

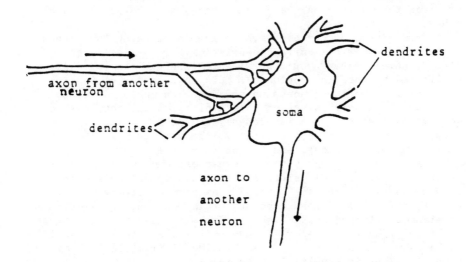

Figure 4. Blueprint of a neuron. The arrows show the
flow of information; the five nodes represent
synapses: switches that are triggered when specific
critical values are reached.

The nervous system has been developed during biological
evolution from fixed and dynamic parts - 'nerves' and
'switches' respectively - and actually it is life's most advanced tool for organic decision making. This leads us to the
broad sense in which 'nerves' and 'switches' will be used. In
the broad sense these terms will refer respectively to the
infrastructure, or 'preformation'[5], and the decision-nodes, or
'switch-formation', inherent in any decision making process in
general. Evidently, every decision making process consists
basically of an infrastructure of data and nexi ('axons') on
the one hand, and of a stock, or possibility space, of con-

struction elements, or operations ('synapses'), on the other
hand. When facing a decision you always have to fall back upon
a structure, supposed to be given: the infrastructure of the
decision. Such a decision making infrastructure encompasses,
for example, the relevant data of the given situation, the
relevant valuations and values (the estimated probability dis-
tributions included), etc. These, of course, are only supposed
to be fixed for the moment; actually, they are part of a dynamic context. Next to this infrastructure you have the choice of
your operations, including the readjustment of your criteria
(i.e. the critical values that trigger a decision in this or
that direction). The two aspects belong together: infrastructure without operations is useless, operations without infrastructure are impossible. In socio-historic reality you meet
with a similar hierarchic complementarity between continuity,
or tradition, on the one hand, and change or revolution, on the
other.

Here nature once again gives evidence of being the mother
of invention whom our current heuristics for the design of conflict control systems would be wise to 'conform' to (this is
explicited further in section 6). This most wonderful instru-
ment for information processing and decision making developed
by nature, the nervous system, has a hierarchical set up which
we have just seen to consist roughly of the following two complementary parts: (1) a preformator, the more or less fixed
infrastructure of axons, and (2) an adaptable informator, the
information processing system of synapses, encompassing memory
as its own traditional part. In a similar way Peirce's category
of secondness encompasses firstness as part of its own.

In what follows, the concepts of revolution, conflict,
complementarity and socio-historic dynamics have been inspired
by biology, Ragazian resurrection-socialism and Peircean social
philosophy[6] rather than by the shackles of Aristotelian philosophy inherent in Hegelianism and Marxism.

3. CONFLICT AND PEACE

In this section conflict is treated in connection with
peace. For the time being let us take it for granted that you
have an idea of peace and that it can be adumbrated as the
"ultimate concern of human community". Let us now turn our
attention to the word 'conflict'. When used in common parlance,
this is a vague and poly-interpretable word; what is more,
valuations of conflict vary from utterly negative to utterly
positive. The mise au point of the vague term 'conflict' is
best begun by elucidating the author's position as concerns
this valuation.

As he conceives of it, conflict is a valuable scarcity. In
spite of the imminent danger of negative effects associated

with conflicts (think of slaughter, etc.), a danger which it is
important to be continuously aware of, I dare say that conflict
has an important positive value in human communication and in
the dynamics of history and society, viz. a positive value in
view of the formation of real community (as opposed to the
pharisaism of a pseudo-community) and therewith in view of the
emergence of emancipation: the liberation from slavery, feudalism, colonialism, and currently the women-liberation movement
encompassing the half of humanity.[7]

In order to explain our stand in this respect, let it
suffice to remind the reader of the well-known device of mental
maturation: "if you never take risks you will never proceed".
This is, of course, not to say that in order to proceed you
always have to take risks. What you always have to do is to
weigh the necessary risks against the intended progress; to
take the example of 'progress by development of nuclear
energy': you can rationally choose not to evolve by means of
this revolution in energy supply when you think that the progress in question is not as substantial as to be worth the
risks in the fields of public health, environment and society
that go with it.[8]

In order to get a clear picture of what we have to say in
this connection it is convenient to radicalise the above device
by substituting it by the following two short formulae:
(1) evolution by revolution
(2) peace by conflict
Being aware that slogans like these can easily be misunder-
stood, by writing them down we consciously take the risk of
being viewed as a simpleton or an extremist. We think the risk
is not too big, since the context in which the two formulae are
presented is clear enough; and we consider the risk worthwhile,
because it leads to a substantial progress in the understanding
of what peace and conflict are like. For a better understanding
let us specify formula (1) and (2) as follows, bringing them in
line with each other:
(1') evolution of society towards a society that incorporates
    more and more the principles of human rights, and – more
    generally – of justice, equity and liberty, feeds on radi-
    cal forces, which as a complement to 'realistic forces'
    refuse to compromise with the prevailing fatalism of in-
    completeness, and which consequently are ready for revo-
    lution.
(2') peace as the apotheosis of communal habit presupposes the
    complete processing of conflict; in other words, peace
    feeds on making conscious, sharing and communalising every
    bit of conflict-stuff that otherwise would latently be
    present as a colourless bar of 'apartheid'.
The fact that slaves have been at peace with slavery for a very
long time seems to contradict what has just been said. This
only appears to be so, because evidently the kind of long-

restrained harmony or 'being at peace' in question has only
barred the way to an ever more integral communal society. The
same holds for every kind of 'slavery'. Thus we arrive at the
following recombination of the two slogans: <u>evolution of
society by emancipation of 'slaves'.</u>[9]

As always, once you start giving specifications, you enter
a self-amplifying mechanism of explanation since every bit of
specification asks for further explanation and specification,
etc. Therefore, we will end this section with a series of remarks which focus on 'peace' and which from there attempt to
throw light on evolution, power, harmony and conflict. Clearly
there is no pretention to completeness.

Peace is not something given, it is not something you can
build on (you have to build your constructions <u>towards</u> it, not
<u>on</u> it), it is not a static situation, let alone a sleeplike
situation of 'pure harmony'. This applies to the negative concept of peace as the absence of war as well as to the positive
concept of peace as the complete communality of men.

The cybernetic anthropology and the dynamics implicit in
both concepts of peace has been characterised in such an exemplary way by Anatol Rapoport at the end of his book on 'Conflict', that it is worth while to quote him here at full
length:

> "The fault is in the consolidation of power, whereby not
> the boundaries of power of the autonomous Self but the
> boundaries of a conquered domain are extended. Uncouple
> the growth of the Self from the growth of power, and the
> contradiction between creative modification of the en-
> vironment and living in peace with nature will disappear
> ..... The crucial question is whether the need for autono-
> my and the appetite for power <u>can</u> be separated, so that
> the former can be nurtured while the latter can be curbed.
> Anthony Storr sees the two drives as the manifestation of
> 'aggression' under different conditions. Aggression (shorn
> of the component of hostility) is, in his view,

> .... *necessary for development, ... for the achievement of
> differentiation ... [but] Competitive aggressiveness ...is
> characteristic of immaturity ... Aggressiveness is at its
> maximum when dependence (and hence inequality) is at its
> maximum; as development proceeds it becomes less important
> till, at the point of maximum development, only so much
> aggression exists as is necessary to maintain the persona-
> lity as a separate entity ...*

> On the other hand, the same author, addressing the problem
> of providing 'substitutes for war', is disturbed by the
> prospect of 'millenary pacifism'.

> *The affluent society cushions us against hunger [and]*

*disease ... and in doing so deprives us of any opportunity
to test ourselves to the limit, to struggle or to die ...
To wait until senile decrepitude puts an end to one's pro-
tracted plush existence is not necessarily an agreeable
prospect.*

No less important than the distinction between autonomy
and power is the distinction between pacific equilibrium
(or millennary pacifism', as Storr calls it) and steady
state .... Open systems (as all living systems are on
whatever level of organisation) are maintained in a steady
state only by constant activity; as Lewis Carroll's Red
Queen says, 'it takes all the running you can do to stay
in the same place ...'. The formidable dilemmas posed by
conflict in man-made environment, whether of man against
man or of man against nature, are rooted in the widespread
failure to distinguish between autonomy and power. By
denying autonomy to others - human beings or even to other
living things (on the scale of the entire biosphere) - the
unbridled appetite for power strangles what is human.
Therefore, if Man's uniqueness is embodied in his awareness of his autonomy as the most precious gift of evolution, men must address themselves to the task of dismantling the fortress of entrenched power."[10]

The motto of the Dutch Queens of Orange: "je maintiendrai"
(I will maintain), as seen from this 'cybernetic' (i.e. governor's) angle, has the flavour of dynamic force and has nothing
to do with the conservation of the status quo. For, within the
dynamic context of history, conservation without a progressive
complement is regression.[11]

'Peace' is a seminal concept; the positive criteria for
peace are time- and culture- dependent. What is to be conceived
as peace has to be readjusted continually; still there is no
arbitrariness.

Although peace development has its ups and downs, still
consecutive stages can be indicated which have a cumulative
effect. For example, an important new stage that was entered
during this century, is the stage of world-wide coordination;
this is exemplified by the institution of the United Nations,
of the International Court of Peace, of the World Food and
World Health organisations and of many other world-wide organisations. Of course, these institutions are still feeble and for
the time being there may be more pharisaism involved in them
then real concern; still they form a frame of reference and a
stand from which humanity is never to resign any more. The same
holds for the universal declaration of human rights.

Peace is a dynamic process; it is something you have to be
ready to fight for - if only in a nonviolent way. Therefore the
positive value of conflict has been stressed at the beginning

of this section. 'Harmony' on the other hand, is usually not
associated with such a struggling force as is needed for peace;
therefore 'harmony' will be given a preponderating negative
sense here. Harmony, if it is not the kind of harmony that has
music to it, i.e. harmony by continuous investment of energy,
has just an invalidating effect, leading to demotivation. A
deteriorating process of mental slackening and physical
slackening sets in as soon as there is too much self-contained
confidence, as soon as there is a collective awareness of
having arrived - a sense of 'pre-established harmony'. Maybe
this process is not irreversible, but it is at least difficult
to reverse.[12] Here we meet with a notable asymmetry in reversibility: the process of 'peace formation', the struggle for
communalisation as indicated above, is easily reversed. This is
grimly illustrated by the 1981-1982 events in Poland. This
asymmetry is an index of the direction of 'evolution'. Just as
biochemical and biological evolution is marked by overcoming
the general egalitarian entropytrend inherent in the second law
of thermodynamics, likewise (and complementary to this) socio-
cultural evolution is marked by overcoming the power flower,
i.e. the elitarian trend that makes power come to whom power
possesses (cf. note 11), inherent in the socio-dynamics of
human power. The 'evil' inherent in the first trend is apparent
to everyone who studies physics, the evil in the latter trend
is clarified in the terminology of Ragaz as Vergewaltigung
('outrage').[13]

4. FROM CONFLICT SITUATION TO CRISIS

We will now try to define 'conflict' in relation to a
series of cognate concepts. Conflict is conceived here primarily as a possible outcome of a conflict situation; the latter
notion is important in view of the fact that we have to deal
here with the objectively identifiable side of a contrapoint
marking the forebodings of a transition from a given status quo
to a new line of development.

Abstracted from details that differ for the many different
applications of the concept, a conflict situation is a situation in which different persons or groups of persons (organised
as institutions, nations, etc.), the parties of the conflict
situation, are operationally and radically at cross-purposes,
i.e. are engaged in clashing interests that have existential
roots.[14] The interests in question may be of any kind:
interests in the vital sphere of food supply, housing, territory, etc; in the economic sphere of energy supply, adequate
infrastructure, supply of raw materials, etc.; in the mental
sphere of privacy, education, schooling, etc.; or in the ideological sphere of freedom, democracy, social wellfare, etc. The
existential impact or radical aspect of a conflict situation

means that the interests in question are actually clashing in such a way that there is no room for all parties to satisfy each one its own interests <u>adequately in due time</u>. As a matter of fact, all analytical problems inherent in the concept of a conflict situation are subsumed in those four words 'adequately in due time'; we will return to it after a preliminary remark. The radical aspect just mentioned rules out those situations of clashing interests where there is no vital scarcity, and which – consequently – go without engagement. Such a situation of clashing interests that lacks engagement rooted in reality does not lead to any kind of enduring conflict, at the utmost it leads to outbursts of emotions, aggression and other symptons of frustration and irritation. Of course, in these latter cases we have conflict-like situations. But what happens here is direct consumption of conflict energy (the emotions are worked off in a flash) instead of accumulated consummation of it, and thus it does not lead to something new, to something outside the sphere of emotions.

Once the parties are aware of their interests and of their competence to satisfy these interests (here are some implict presuppositions which will be made explicit below), it is plausible to suppose that each party tries to satisfy its interests at least 'adequately in due time'. At this point it is crucial to discriminate between two concepts of 'adequate satisfaction in due time'; these two concepts involve two completely different concepts of (economic) decision theory, the first of which reduces conflict theory to game theory.

Concept nr. 1. is affiliated to (neo-)classical economics; it incorporates a nonsocial or maximal approach to 'adequate satisfaction in due time' which can be phrased as: "sweep in as much as you possibly can" or "select those operational alternatives that yield statistically and combinatorily the best results with respect to your proper interests."

This concept transforms conflict (situations) into game (situations); there is no proper conflict, because the radical aspect is not vitally rooted in existence. It corresponds to the type of rationality which has become current in the Western world and which could be characterised as non-existential rationality. Because of its enormous impact in the Western way of life and in the economy of 'free' markets and competition, this approach instigates institutions such as shops, multinationals, nations and even universities, public health service and social services to play the competitive optimisation game of 'never enough', whether they like it or not.

By this approach, the social aspect inherent in economy, policy, etc. is greatly reduced to coalition, which in turn is conceived operationally in game-theoretical terms. The well-known 'monopoly game' simulates this state of affairs in a very simple but illustrative way. What is wrong with the non-exis-

tential rationality in question, is that it comes down to a
'nowhere' concept of (the effects of) power, the complement of
which is (implict or explicit) power-worship.

Concept nr. 2. interprets 'adequate satisfaction in due
time' in a communal or minimal way; this leads to a sufficiency
economics as promoted here in Holland for instance by Goudzwaard.[15]) This minimal attitude entails vital engagement in
clashing interests. Thus, here we meet with real conflict situations instead of just competitive game situations; at the same
time, however, and this is an important supplement, this attitude brings with it a sensibility for reciprocity and a readi-
ness for sharing whatever turns out to be vitally scarce. This
considerably reduces the risks of conflicts getting out of
hand. Sharing in this context is the positive version of compromise. What is 'adequate' and 'sufficient' has to be constantly redefined in the concrete social situation; this means
that conflict is met here primarily in the positive form of
confrontation.[16]) Likewise, we have here cooperation instead of
coalition; whenever cooperation is rejected by the other party
it has to be obtained by confrontation; when confrontation
fails, the only way out is conflict. As usual, reconstruction
of this conflict concept in game theoretical terms is possible
in principle; it would lead to a dynamic version of game theory
in which the positions in the game, the balance of power, and
the definition of sufficiency are not fixed but are dynamically
related to each other and to the rules of the game that change
after each confrontation, c.q. conflict[17]), and this leads to
bifurcations that keep restructuring the game like a kaleidoscope. But different from the usual kaleidoscope, the consecutive patterns exhibit a progressive line.[18])

It should be noted that H.A. Simon's well-known 'satisfi-
cing' approach to decision making, which has become classic in
the fields of artificial intelligence and heuristics, bridges
the concepts nr. 1 and nr. 2. Another way of bridging them goes
by the concept of 'bargaining room'. If there is no bargaining
room, then the two concepts more or less coincide and conflict
is reduced to a zero-sum game. This is, however, an artificial
and unrealistic case, and when it applies it is usually due to
the blindness of the parties for lateral thinking and acting
(cf. note 4). When there is some bargaining room, i.e. some
common benefit (some parallel interests next to the clashing
interests) then there is a restricted operational basis for
coalition, c.q. cooperation. The two concepts then give rise to
two strongly differing attitudes:
-    attitude nr. 1: the social aspect is not accepted unless
     the common benefit can be proved (the hawk), and as far as
     it is accepted it is treated in a tricky way (the guileful
     serpent);
-    attitude nr. 2: the social aspect is welcomed a priori
     until it is proved to fail (the dove), and it is made use

of in a maximal way (the communal bee).

Given a conflict situation, then after a gestation process that ends in a confrontation, one of three things eventually emerges: (1) a conflict, (2) a disguise, or (3) a transformation of the conflict situation. The latter possibility, which will be given no further attention here, may consist of the attainment of a compromise without disguise, that is a compromise in which each and all parties engage without reserve, without resent[19], of the healing of the situation without there first being an outburst of conflict (for instance by intervention from outside), or of any other redefinition of the conflict situation. The second possibility, the disguise, means a delay of conflict – maybe for a very long time – because the conflict-stuff is still there. Usually such a disguise or delay has only a negative effect; the effect being a delayed conflict with substantially higher risks as concerns outbursts of emotions, etc. that have a way of getting out of control.

What happens when the outcome is a conflict? If the conflict goes the whole hog and is not interrupted in its way out by the realisation – afterwards – of possibility (2) or (3), then it will ultimately lead to a crisis. Here again we meet with a notion that is of central importance, because it denotes the stage where the forebodings of the contrapoint have arrived at a point (better: a funnel) of no return irresistably leading to a decisive bifurcation.[20] From the very moment that a crisis sets in one thing is sure: the situation will never again be as it has been before: it will either be better or worse. This is what a psychiatrist told me in connection with mental crisis, this is what prime minister Den Uyl declared at the beginning of the oil crisis (1973), and this is what is generally known about the critical stage of a disease.

Originally 'crisis' was only a medical term. The word 'crisis' stems from the Greek verb κρινειν, which means to winnow, to decide. It denotes a stage in a disease that is decisive for the further course of the disease – usually a question of life or death. By and by the term 'crisis' has got applied to a much wider range of analogous phenomena; such phenomena – pregnant with vital bifurcations – are well-known in the mental, the socio-psychological, the economic, the governmental and political sphere, as well as in the field of development. As for the latter, the outstanding example is the instable period of puberty. This usually covers several critical stages in development: an authority crisis, an identity crisis, etc. What is particularly enlightening in this example is the circumstance that here we have a phenomenon, generally conceived as 'sound, vital and common', which at the same time challenges ordinary views and conformism because of its destabilising conflict- and crisis character.

One important point should be added to the above analysis.

It should be noted that a conflict situation does not by itself
lead to confrontation; at least two stages of maturation are
presupposed in the above analysis. First of all, the 'objective' conflict situation should be interiorised ('subjecti-
vised') by the parties in question. For, why should 'proletari-
ans' confront or unite if there is no (awakening of) conscious-
ness of unjust insufficiency. The latter notion is crucial.
Here exactly we meet the basic motivation: the feeling and
conviction of being ruled out as a 'quantité négligable'.
Secondly, there will be, at most, a confrontation but no real
conflict as long as there is no awareness of the ruling powers
and as long as there is no formation of new power – if only the
power of nonviolent action and solidarity – to oppose these
powers of the status quo.[21]

Once the first stage of maturation has been reached there
is no way back any more. Every form of disguise then creates a
new situation, experienced as schizophrenuous or pharisaistic:
a disguise appears where no disguise can be. It may be helpful
to compare a conflict situation with a dormant volcano. As soon
as the maturation conditions are fulfilled the volcano becomes
active. In this connection, strikes, model-actions, demonstra-
tions, the development of big movements of solidarity and
peace, etc. are like smoke, signalling imminent volcanic outbursts in the near or far future. At this stage several things
may happen. Let us look at four of the many possible scenarios.
The following 'scenarios' are rendered very short; they are
only intended to give an idea of possible processes once a
conflict situation has matured.

Scenario nr. 0, for instance, is a real zero-scenario
where the authorities act as 'nowhere men'; the result being an
autonomous development which leads to an outburst of conflict.
Scenario nr. 1: The authorities give in, and that in due time;
this usually is an ideal crisis-preventing solution, unless we
have to deal here with a 'reversed disguise' of the conflict.[22] Scenario nr. 2: The authorities respond in a polari-
sing way; the result being a quick oscillation of conflicts.
Scenario nr. 3: The imminent conflict is canalised (e.g. mani-
pulated) by the authorities. This is the interesting option
which has several variants. Here, again, we have the possibility of disguise, which is the least interesting variant. The
most interesting variant here, for which we ask special attention, is what we like to call 'constructive canalisation': the
conflict is controlled in such a way that the imminent crisis
has a fair chance of not being left to random outbursts of
irrationality and mass-psychotism only. This may involve the
installation of a crisis-centre, for example, etc.

The ultimate result of a conflict is the bifurcation that
is usually reached through a crisis. The whole range of alternative outcomes, resulting in the 'critical bifurcation' in
question, may be represented by the following three features

(cf. Figures 2 and 3):
1)   a new synthesis, where the conflict stuff is fully
      digested ('evolution'),
2)   a disguise, c.q. delay, of the problem (a provisional fall
      back), and
3)   a dictatorship of destructive forces, of which war is the
      worst example (a definite fall back).

The dynamism inherent in bifurcation implies that shortly
after the consummation of the crisis bounds from semi-stable
alternative (1) to semi-stable alternative (2) or (3), and vice
versa (and to and from the intermediate alternatives) are quite
possible.

5. CONFLICT CONTROL SYSTEMS

In nature and technology the development of control systems is motivated by constraints c.q. norms: the value of certain variables has to be maintained between an upper and a
lower bound, the critical limits of the variable in question.
Whenever such a critical limit is passed, the subject or information processing system to which the control system is applied
enters a crisis and may be bound to die. Think, for instance,
of the control systems regulating the sugar level in the blood
and the temperature of the body. Usually the control system is
serving an information processing system and is itself a subsystem of it. What it comes to is that updated information
about the median-deviance from the norm is fed back by way of
steering signals (triggered by the regulator) to the information system which in this way adapts to its norm.

We now present some preliminary remarks about the design
and use of conflict control systems. The first thing to do is
to reconstruct a survey of what is tolerable c.q. sufficient,
and what is not. This is culture- and situation dependent.
According to the principle of management by exception only the
non-tolerable and the insufficient are taken into consideration. Further, a survey is needed of to know what kind of
variable can be regulated by what kind of mechanism. On the
basis of this and other information it is possible to design
conflict control systems which can handle their task. Such a
system is open in two senses: (1) as an information processing
system, and (2) as a nontotalitarian system, i.e. as a system
whose programme and operation are continually subject to even-
tual adaptation and transformation due to decisions from the
side of the governor or institution that runs the control system.

This brings us to some further issues. The actual design
and use of a conflict control system depends to a considerable
degree on the answer to the following questions: (1) for which
institution (c.q. in the hands of whom) is it supposed to oper-

ate? And (2) at which stage of conflict is it supposed to be
operated?

The second question is easily answered. Conflict control
systems are designed in order to prevent conflicts from getting
out of hand. They are, say, 'war' preventing and certainly not
conflict preventing. Dictatorship is a simple example of a conflict preventing control system as well as a clear illustration
of its drawbacks. Thus, it is at the stage of the crisis, or
just before it, that a conflict control system should be made
to operate.

Question nr. 1 is crucial and problematic. For, the natural operators of such a control system are the authorities, i.e.
those who usually back the vested interests of a majority which
are compromised by the very conflict. The majority in question
may be qualified as a majority in sheer number, or in social
position, or in property, knowledge, or any other socio-economic dimension of power. It should be noted that even in ideal
democracies the relevant authorities have been given an ambi-
valent competence, the negative side of which can be perceived
as a benevolent – but still repressive – dictatorship of majorities. Endenburg has convincingly indicated how minorities have
an insufficient role to play in democratic societies, and how
this insufficiency can be overcome in sociocratic societies[23],
without giving pre-eminent minorities an overruling power (as
is usually the case in communist and militarist societies).
This means that as long as we do not live in a sociocracy, the
authorities are party in a conflict. This is a pity because
they would do a good service operating conflict control systems. But, according to a well-known principle, no party can
justifiably be given the arbiter's competence to operate a
conflict control system. Actually, as everybody knows, the
authorities are still given that competence. And often, as
things are, this is the best way to do the job.

A case in point is the way in which jurisdiction is regulated in modern society. This is a well-balanced conflict control system operated by professionals (sometimes counterbalanced by laymen) to whom this competence is delegated. The character of being an independent institution is the important
issue here. As concerns this best of all institutions it should
be noted that it is too much of a self-proceeding system as
long as it has not been embedded in a sociocracy. As a consequence there is an enormous distance between jurisdiction and
human experience; a distance which it is impossible not to
stuff with whatever turns out to parallel the interests of
majority powers. This makes even the best of all systems a
class bounded, or at any rate a majority bounded, system
whether or not those handling the system like it.

As to international conflicts, the principle that excludes
parties from operating conflict control systems is, of course,
of special significance. This asks for independent institutions

on a united-nations level that are given the power to act as
third party in the conflict and to operate any conflict control
system that will turn out to be judged adequate and justified.
And this in a future which may not be tolerated to stay future
for ever and ever. The problem then arises along what lines of
political organisation and decision making such a jury power is
developed and such a judgement about the adequacy and justifi-
ability of control systems is legitimated. As we see it at the
moment, sociocracy is the answer. Until now sociocracy has been
tried out and tested for decision making organisations on a
micro and a meso level. As such it has proven to be a sufficient conflict control and conflict solving system within the
given context. There is still an enormous task before us: to
design sociocratic organisations, to specify new sociocratic
principles that do the job for ever more comprehensive organisations, and to realise them juridically and politically,
therewith stemming the tide of the socio-dynamics of power
indicated above as power flower.

　　In order to be careful not to get biased into a form of
'idiosociocracy' it is good to recollect U Thant's "five steps
for world change" (1971), the third of which reads:
　　　　"Claims to exclusiveness of political and social
　　　　systems (that is, of exclusive possession of truth,
　　　　principles of social justice, etc.) must be abandoned.
　　　　'No rigid system, however well established on ...
　　　　principles, is able to cope with ... constantly
　　　　changing society'."

Rapoport (l.c, p. 241 f), from whom we just quoted U Thant,
makes the following comment in this connection:

　　　　"Prescriptions of this sort ... entail a radical
　　　　change in the perception of values, ... and social
　　　　roles. The perceptions of people in positions of power
　　　　and of those who serve power in professional
　　　　capacities are especially difficult to change, not
　　　　only because the prerogatives of power are not easily
　　　　relinquished but also because people in power are
　　　　often convinced that in exercising power they are
　　　　discharging awesome responsibilities."

6. THE NERVES OF GOVERNMENT AND THE SENSIBILISATION OF CULTURE

　　What the above comes to may be summarised on the basis of
the following two propositions which we will briefly comment on
in this section:
1)　　Conflict control systems have a 'fixed' and an 'open' part
　　　which can be characterised as the system's infrastructural
　　　set up or tradition (preformation) and its adaptive space

or developmental capacity (switch formation).
2)   It is not the imminent conflict, c.q. crisis, as such that
     has to be prevented with the help of conflict control sys-
     tems but those of its possible effects that are 'war-
     loaden'.
The structure of this final section is determined by the
complementarity of 'nerves' and 'switches' in the context of
'war' prevention. This complementarity corresponds to the complementarity of realism and idealism in politics, as follows:
realism that starts from the status quo, from factual information as preformation, should be complemented by some kind of
constructive idealism that utilises switch-formation for the
realisation of human community.[24)]

6.1. Realism

     In the course of evolution one thing happened that is of
major importance for a proper understanding of realism: the
struggle for life evolved into the communication of conflict.[25)]
To call the competition for survival between selfish genes a
matter of conflict would be too metaphorical. Only at a stage
where such a competition can be said to be partly mediated by
some specialised means of communication (signal 'languages',
gestures, displays, etc.) between organisms or groups of organisms that embody a selfish gene the term 'conflict' can be
meaningfully used to describe what is going on. From the point
of view of evolution then communication should be conceived in
terms of a control system: endogenous communication for self-
control, exogenous communication for social control. One of the
main control functions of communication (especially of exogenous communication) is conflict control or more precisely: the
expression and the control of conflicts due to clashing
interests. The highest developed tool in this respect, developed in the course of evolution for the purpose of communication,
control and decision making is the nervous system together with
facial expression (gestures, vocalisation, etc.).[26)] 'Nerves'
and 'switches' - in the narrow and in the broad sense - form
the fundamental pattern of the nervous system as well as of
communication and control. All this is illustrated in an excellent way for the case of endogenous self-control of a flock and
for the case of exogenous interflock conflict control by the
following two descriptions of the communicative behaviour of
the so-called 'howling monkey' by Southwick:[27)]

          "Among the most notable features of their social
          behaviour ... are their famous patterns of vocali-
          sation ... which integrate group life, control group
          movements, maintain social positions within groups,
          and regulate spatial patterns between groups."

          "The rhesus seemed to have no behavioural mechanism

for avoiding ... harmful conflicts. They had no
ritualised patterns of vocalisation that enabled
groups to avoid each other without visual contact. In
contrast to this, howler groups ... have ... effective
intergroup communication at the auditory level ... If
two groups ... come into close proximity they engage
in ... 'howling battles' ... There is obvious
intergroup antagonism, but not direct fighting occurs
... There seems to be a strong behavioural inhibition
against direct fighting."

As far as can be spoken of a 'monkey-culture', the
'howler-culture' has apparently attained a higher degree of
sensibility than the 'culture' effectuated by the communication
among rhesus monkeys: actual war is so to speak substituted by
symbolic war. As seen from this perspective, the original question put forward in the introduction all of a sudden appears
rhetoric. Conflict control systems that are less menacing than
the outbursts they try to prevent are widespread: communication, symbolisation and sensibilisation are the answer. The
(in)human history of warfare is a kind of upside-down evolution: it has become more and more harmful. Our generation goes
through one of those Sternstunden der Menschheit inevitably
leading to a bifurcation: doomsday or retreat from evil[28]),
i.e. deflower of power by the sensibilisation of culture (evolution).

And here we are exactly at the main issue of what we have
called 'realism'. The incredible 'paradox' inherent in conflict
control systems being more menacing than the outbursts they
intend to prevent is not due to some paradox character of reality as dialecticians ever want us to see it, but simply to a
pharisaistic ambiguity in the intentions of the powerful in
East and West. They want the impossible: they want to serve
peace (as the official military ideology tells us in East and
West) and power (as realism and actuality tell us). Some of the
macro-level conflict control (sub)systems, e.g. diplomacy,
balance of terror, military force and sometimes - namely in
pseudodemocracies - also police force have an official function
which is not in line with the actual function they have simultaneously. Officially they are intended to prevent those
possible effects of conflicts that are 'war-loaden': they have
to prevent the exogenous or endogenous world from becoming a
jungle where only the club law counts. Actually, however, they
are in no lesser degree at the service of power flower - as
every 'third party', especially the tiers monde, can easily
ascertain. This improper use has been given its classic
expression in Clausewitz's formula: war is the extension of
diplomacy with nondiplomatic means. According to Clausewitzian
realism it would be self-contradictrory to call the afore mentioned macro-level systems 'conflict control systems': they are
conflict escalation systems at the service of power, influence

and growth. This is illustrated by the 'paradox' that according
to the official version the outburst of war would mean the
absolute failure of diplomacy and full reason for the responsible authorities to send in their papers, whereas what usually
happens is a fortification of their positions.

6.2. Idealism

The future of the world will be a bifurcational future of
peace. The crucial constructive issue is whether it be a peace
of the cemetery or a peace of living community. As to the outcome of this imminent vital bifurcation (vital because one of
the two 'solutions' at the imminent singular point has the
value zero), much will depend upon the creative struggling
force of peace movements all over the world triggering at least
two new developments that are complementary:
1) The repropriation of conflict
2) The sensibilisation of culture.
We have coined the word 'repropriation' after reading B.
Moore's "Expropriation of Moral Outrage" (l.c., section 14.7).
This subject is treated by Moore in analogy to the expropriation of the means of production, and the expropriation of guilt
as effectuated over the centuries by the Catholic Church: "It
has achieved this by helping to create the sense of guilt and
then providing the bureaucratic mechanisms for alleviating it
... The Catholic Church managed to create much of the demand
and most of the supply". Next, Moore explains how moral feelings and expression are expropriated in modern society (the
role of the media, etc.). Now there is one condition which is
essential for conflict having the capacity to play the evolutionary role as it was depicted in section 3: conflict should
be authentic. For conflict to be authentic, two conditions that
are narrowly related should be fulfilled: the repropriation of
moral feelings, of (self)expression, of valuation, etc. and the
primacy of self-criticism or endogenous conflict above exogenous conflict. Explanation of the latter deserves a special
lecture; the first can be summarised as repropriation of the
self, something which may be realised by not shunning endogenous conflicts.
As this process of repropriation sets in and proceeds,
conflict stuff will be expressed, digested, and solved progressively in a symbolic and existential way: communication and
confrontation will substitute war and conflict. The apocalyptic
key, it seems to us (after reading Levinas), is sensibility for
the human face, i.e. for what is vitally human.
Remembering the many many persons who are tortured or done
away with as a quantité négligable all over the world at this
very moment, we realise how far we still have to go.

NOTES AND REFERENCES

1) The present version differs considerably from the original
version presented during the symposium. This comes from a
bifurcation in career that happened just before the symposium was held, and that has changed my professional life to
a considerable degree. Staying professionally in the field
of philosophy, I switched from logic to peace research.
This, actually, is my first public appearance as a peace
philosopher. Notwithstanding all the revisions done, the
present text – I am sure – has still its shortcomings.
'Peace philosophy' is still greatly a blank that is yet to
be filled. Two of the 'seven pillars of wisdom' needed for
peace philosohpy will be – I think – interdisciplinary
cooperation and a complete conversion of philosophy. An
exemplary vision of the latter has been put forward by
Emmanuel Levinas. In the preface to his Totalité et Infini
(Nijhoff, The Hague, 1971⁴) he confronts the totalitari-
anism of traditional Western philosophy with a practical
way of philosophising that attaches primacy not to ontology
or foundations, but to the open face of peace and the
infinite way of community that goes with it.

2) See for instance C. Hartshorne and P. Weiss (ed.), (1960²)
Collected Papers of Charles S. Peirce, Cambridge/Mass.,
Vol. 1, sections 347 and 363; Vol. 3, sections 468 to 491.
See also C. Eisele's edition of Peirce's The New Elements
of Mathematics, Mouton, The Hague, (1976), Vol. 4, Chapter
17.

3) For Peirce's theory of categories and its relation to
evolutionism, communalism, ethics, self-sacrifice, self-
control, cybernetics and semiotic anthropology (man viewed
as a sign) see references in note 2, and Collected Papers,
Vol. 5, sections 442 and 533 f; Vol. 8, section 320.
Further K.L. Ketner et al (ed), (1981), Proceedings of the
C.S. Peirce Bicentennial International Congress, Texas Tech
University, Lubbock, Texas, especially the contributions
of: M. Ayim, R.J. Bernstein, D. Nauta jr., S.M. Harrison,
E.R. Tarr and M. Nadin. See also J.K. Feibleman, (1970), An
Introduction to the Philosophy of Charles S. Peirce, MIT
Press, especially pp. 382-386: "The unlimited community".

4) As far as this principle applies to (methodo)logical
systems, it has been explicited by E. de Bono (The use of
lateral thinking, (1967)). As far as it applies to juridi-
cal laws, it has been explicited by legal philosophers. And
as far as it applies to ethico-religious rules and rituals,
it has been explicited by the rabbi who gave us 'sabba-

thical leave'. See for man's ambiguous position and excentricity Peirce (note 3) and H. Plessner, (1928), Die Stufen
des Organischen und der Mensch, Berlin.

5) For preformation see C.A. van Peursen, C.P. Bertels and D.
Nauta (1968), Informatie, Het Spectrum, Utrecht, p. 82 and
p. 84. The term indicates the prefabricated structure presupposed for information to be meaningful. Think of genetic
codes (DNA), frames of reference, and generally the information relevant structure of an information processing
system. As to infrastructure, the Oxford Illustrated
Dictionary tells us that it designates the "supporting
system of any organisation", especially "the fixed instal-
lations and facilities necessary to support military
operations, as airfields, naval bases, training establish-
ments, supply works, etc.".

6) Leonhard Ragaz, (1936), Sinn und Werden der Religiös-
Sozialen Bewegung, Zürich. Peirce, see note 3. Further J.
Royce (1968²), The Problem of Christianity, University of
Chicago Press, which according to D. Greenlee ((1973),
Peirce's concept of sign, Mouton, p. 98) is "the first
serious philosophical elaboration of Peirce's semiotic".

7) According to H. Hartmann ((1964), Essays on Ego-Psychology,
International Universities Press, New York, p. 12) conflict
is a basic ingredient of development. Similarly, according
to A. Storr ((1963), The integrity of the personality,
Penguin Books, p. 81) frustration is essential for the
maturation of self-hood.

8) Throughout this paper evolution (progression or development: vital, mental, social, etc.) is taken as a positive
breaking process in which every stage embraces the preceding stage and specifies it further. In this sense
Einsteinian mechanics is an evolution of Newtonian
mechanics, sociocratic society is an evolution of democratic society (see G. Endenburg, (1981), Sociocratie,
Samson, p. 39), etc. In this sense massive deployment of
nuclear energy is a technological evolution. However, the
idea that this does also mean an integral evolution, i.e. a
societal progression, is seriously questioned by Robert
Jungk ((1977²), Der Atomstaat: vom Fortschritt in die
Unmenschlichkeit, Kindler, München). Atomic terrorism is
one of the issues here which directly relate to our
subject.

9) Marxist explanations of these matters are usually very
simplistic, the more so as they get to be called 'scientific'. A valuable sourcebook indicating the many squabbles

that make social evolution a nonlinear process is B. Moore
jr., (1978), Injustice: the social bases of obedience and
revolt, Sharpe.

10) A. Rapoport, (1974), Conflict in man-made environment,
Penguin Books, p. 249 f; A. Storr (l.c. (note 7), p.56 f,
and A. Storr, Possible subsitutes for war, p. 141 in J.D.
Carthy and F.J. Ebling (ed.), (1964), The natural history
of aggression, Academic Press. What is called here "man's
unique (awareness of) autonomy" by Rapoport is exactly the
same what was meant by the excentricity, creativity and
responsibility of man (cf. note 4).

11) It seems to us that there is one thing on which conserva-
tives, liberals, communists, bourgeois socialists and
fascists agree at this point of 'steady-state policy', viz.
the serious misconception that progression in power,
specified as surplus armament, surplus police force, surplus economy or surplus standard of living, is such a progressive ingredient that cannot be missed in such a policy.
The effect is that these powers become a self-continuing
and self-growing institution: the keen spirit of 'the
creative animal' is in danger of being stifled by the
unwieldiness of the cancriform mammoth-beast. For an
alternative, prophetic, vision of 'je maintiendrai' see
Jesaiah 42:1.

12) Note that Wagnerian hero-worship, Prussian militarism and
Germanic idealism, which form the complement to fatalism
and which are spread nowadays over the two thirds of the
world that are under control of capitalism or communism,
(in this sense it is certainly true that "Hitler won his
war" as is explicited by M.J. Top in the paragraph under
this title, in his book Leonhard Ragaz - religieus socia-
lisme contra nationaal socialisms, Kok, Kampen (1977))
lead to a similar semi-irreversibility: they introduce a
positive feedback of surplus power without negative controls. War and harmony appear to us here as the two unwork-
able extremes, the scylla and the charibdys between which
you have to sail using the complementarity of peace and
conflict as your compass. K.H. Miskotte, (1939), Edda en
Thora, (Callenbach, Nijkerk), should be mentioned in this
context as an eye-opener as concerns Germanic fatalism and
its Jesaian alternative.

13) This applies to maintenance of the status quo in as much as
it just parallels power flower. Self-reliance of powerless
individuals or groups is then 'vergewaltigt', overruled, by
those in charge of power. A hard core criterium for (movements of) peace is: freedom from power-worship. This is the

small way of the few and the happy poor, as opposed to the
highway of majorities leading to the Kingdom of Power. Of
course, power is not a filthy thing – if only it be used
for 'evolution', i.e. to combat power flower. For Ragaz,
see note 6.

14) Apart from the addition of a radical c.q. existential
element, this definition agrees with those given by J.
Galtung, (1969), Violence, Peace and Peace Research,
Journal of Peace Research, p. 167ff., H. Schmid, (1968)
Peace Research and Politics, Journal of Peace Research, p.
217ff. and many others. The addition of a radical element
here, if taken seriously, means a priorisation of concept
nr. 2 below. In case of what would be called an endogenous
conflict situation by Rapoport (l.c., p. 175), a specification of the definition applies: there is only one person
or institution, subdivided into different subpersonae or
departments that are the parties of the conflict situation.
Subpersonae conflict psychology applies to sane as well as
insane constellations ('knots'). Examples of the former are
the alter-ego or conscience critisising the ego in matters
of morality, scientific truthfulness, etc. Examples of the
latter are non-communication in certain academic circles
because of a repressed bad conscience (in matters of scientific truthfulness), the emergence of anguish and phobia
because of subpersonal repression, schizophrenia, etc.

15) See for instance B. Goudzwaard, (1978^2), Kapitalisme en
Vooruitgang, Van Gorcum, Assen, and (1974), Schaduwen van
het groeigeloof, Kampen.

16) This key word stems from the Latin frons, which means face.
Accordingly, confrontation is understood here as profiling
the existential issues of clashing interests by revealing
them face to face. Here we meet with the Levinasian order
of peace: the open face as ground or witness from which
truth emerges. Levinas (l.c., see note 1) opposes this to
the current ontological order according to which truth is
the pre-given factum from which acts should be derived.

17) The kind of game you get would be more like simulation
games as treated for instance by P. Boskma and F.B. van der
Meer, (1974) in: Simulaties van Internationale Betrekkin-
gen, Tjeenk Willink, Groningen than like games as they are
treated mathematically in the tradition of game theory. It
should be added, however, that the simulation games treated
by Boskma and Van der Meer lack the sensitivity aspect that
is substantial for confrontation and for the kaleidoscopic
effects on the restructuring of the game (l.c., p. 82,
where lack of the sensitivity aspect is connected with

fixed structures as the hard core of the game).

18) Cf. the development of science as conceived in terms of the perspectivist Gestalt-theory put forward by C. Dilworth, (1981) in Scientific progress, Reidel, Dordrecht. According to this concept successive scientific theories "are often related in the same sort of way as are the different aspects of a gestalt switch diagram. Following this lead, a model ... is introduced ... intended to ... afford notions both of conflict and of progress." (l.c., p. 112).

19) This is what sharing comes down to: resolution by communal decision making, which is the hard core of sociocracy as put forward by Endenburg (see note 8). This outcome of the conflict situation corresponds to what in Figure 1 is rendered as "new synthesis'; it is actually a pre-conflict version of it.

20) In 'sensitivist societies', i.e. societies where sensibility or fraternity plays an overruling role, the same 'funnel-effect' is met already at the stage of confrontation, i.e. before a conflict in the more radical sense is emerging (cf. note 19). As seen from this perspective, other kinds of societies are conflict provoking. In order of degree of provocation, complemented with repression, three kinds of societies can be distinguished here: 'utilist', 'institutionalist' and 'militarist' societies.

21) Throughout the paper the focus of interest is asymmetric conflict from the point of view of the 'opppressed'; and this in the context of concept c.q. attitude nr.2. What matters here is not conflict for the sake of conflict as has been appraised in Germanic philosophy (finding its apothesis in Steinmetz' worship of war), but the utilisation of confrontation in a peaceful way. Therefore, the mobilisation of new power should not itself have the character of what it intends to oppose: power flower. Instead, it should have the character of 'energy for peace': a kind of flower power that has the guts to maintain the powerless.

22) Disguise is characterised by a denial of interests. Usually, the party that is the most powerless gives in and acts as if its interests are not denied. In case of a reversed disguise the powerful party, for the time being and resentingly, gives in. Usually this is a choice consciously done in order to be able to reverse the situation later on.

23) Of course, parliamentary control of governmental power is a strong case in point for democracy. There are at least two

serious shortcomings, however, which finally will put an
end to this form of democracy. The control in question is
an institution that resides too far from the people whom it
concerns. As a consequence, the parliamentary output testi-
fies of a serious estrangement from the basis. Secondly,
there is no way in which autonomous individuals or groups
gain their say in matters that concern them vitally, i.e.
democracy is fundamentally deficient as an input for the
voice of minorities. The democratic control system is far
from integrated, and additions like business councils have
the character of a proviso only. Endenburg (see note 8) has
indicated how sociocracy as the successor of democracy
brings power control on a level with the individual whom it
concerns, without being constrained to a constant delibera-
tion with everyone on everything. This is - as it seems to
us - more relevant, constructive and 'revolutionary' than
anything put forward by Marcuse and others since the 'big
Russian revolution'. Sociocracy can be characrterised as
'repropriation of power and power control' (cf. end of last
section).

24) Throughout the paper the hierarchy of complementarity is
defined in such a way that the 'higher' depends on the
'lower' in the hierarchy. Josiah Royce (l.c., p. 32 f)
defines community as mediation of conflicts, and as the
only way to overcome "the twin evils of individualism and
collectivism".

25) This is how Richard Dawkins describes the process of
evolution in his book The Selfish Gene, Oxford University
Press, 1976.

26) The pre-eminent role of the face, its capacity for
expression and for reciprocal recognition, in the evolution
of communication is stressed by Levinas and by many biologists, e.g. Dawkins (l.c., Chapter 10) and J. van Hoof who
specialised on the subject.

27) C.H. Southwick, Challenging aspects of the behavioral
ecology of howling monkeys, in C.H. Southwick (ed), (1963),
Primate Social Behavior, van Nostrand, p. 186 f.

28) This subject is treated from a christian apocalyptic point
of view by B. Goudzwaard in his last monograph, entitled
'necessitated to be good' (Genoodzaakt goed te wezen,
Kampen, 1982).

BIFURCATION AS A MODEL OF DESCRIPTION - A MEANS OF MAKING THE
HISTORIOGRAPHY OF PHILOSOPHY MORE HISTORICAL?

Frank Vleeskens

Central Interfaculty
Erasmus University, Rotterdam

1. INTRODUCTION

Historians of philosophy are a kind of go-between. It is
their task to bring about a relation between two disciplines -
history and philosophy. It is thus no wonder that the complaints
about them have come from both sides. An example of the complaints from the side of philosophers may be found in the Preface to Kant's Prolegomena. In his opinion, historians of philosophy tend to overemphasise the resemblances between contemporary philosophical ideas and philosophical ideas from the
past. Kant's suggestion is that by doing so they discourage
philosophers. Therefore they "must wait until those who attempt
to draw from the fountain of reason itself (i.e. the philosophers, F.V.) have made their case; it is then their task to
tell the world what has been done".
    In this paper, however, we would like to concentrate on an
example of the complaints about historians of philosophy from
the side of historians. In his article "The idea of a history
of philosophy", John Passmore (1965) argues that nearly all the
ways in which historians of philosophy have approached and
still are approaching philosophers from the past are unaccept-
able from the point of view of historiography. The first part
of our paper will be dedicated to an analysis and evaluation of
Passmore's opinion. Especially his pleading for the so-called
problematic approach gives a clue, we think, for a positive
answer to the question we shall be concerned with in the second
part: whether bifurcation as a model of description might be
useful to make the historiography of philosophy more historical, i.e. less incompatible with what we know about the history
of philosophy.

M. Hazewinkel et al. (eds.), Bifurcation Analysis, 141–147.
© *1985 by D. Reidel Publishing Company.*

2. PASSMORE'S OPINION ABOUT HISTORIANS OF PHILOSOPHY

Passmore's article is designed as a reply to Kant's complaint about historians of philosophy in the Preface to the
Prolegomena - a complaint which in Passmore's opinion sums up
the attitude of a great many philosophers to the historians. At
the same time his article provides an anlysis of the methods
used by historians of philosophy and contains a proposal for a
better, more historical method.

There is some truth in Kant's complaint, Passmore says,
because indeed some historians of philosophy take a delight in
finding 'precedents' and refuse to admit that there is anything
fresh to be said. On the other hand, this attitude is not
necessarily useless to philosophers. This does not mean, however, that everything is all right with the historiography of
philosophy. In his analysis of the ways in which historians of
philosophy have been dealing with the history of philosophy and
still are dealing with it, Passmore succeeds in showing that a
considerable part of the work done by historians of philosophy
has to be qualified as 'bad historiography' or even 'no historiography at all'. He draws a distinction between three
different ways of approaching philosophers from the past: the
polemical, the cultural and the elucidatory way. In his evaluation of these three ways the main criterion is the degree of
historical sense exposed.

Polemical studies about philosophers from the past, as
Passmore points out, "arise out of the tradition that philosophy ought to be taught by way of criticism of acknowledged
masterpieces rather than by way of introductory manuals" (p.7).
Philosophical positions are defended or attacked themselves, or
are used in defending or attacking other positions, without any
attention being paid to their historical setting. In Passmore's
opinion it is misleading to present polemical studies under the
heading of history of philosophy, because they rather are
introductions to the several disciplines of philosophy. In
cultural studies philosophical theories are exclusively considered as representative expressions of a period. Passmore
rejects this approach; in the first place because it disturbs
"the balance required of the intellectual historian .... to
emphasise at once the historical roots and the permanent
importance of a contribution to philosophy ...." (p.15), and
secondly, because in this approach the vertical or internal
relations within the history of philosophy (i.e. the ways in
which philosophers are influenced by their predecessors) are
neglected.

Within the elucidatory studies "which are concerned to
tell what actually happened within philosophy" (p.18), Passmore
makes a distinction between the doxographical, the retrospective and the problematic type. "In a doxographical history the
major conclusion of philosophers are concisely set out, gener-

ally to the accompaniment of a certain amount of biographical
detail. The chronological framework is provided by the idea of
a 'succession', or a school, alleged to indicate a line of
influence ...." (p.19). The main difficulty with studies of
this kind is that they concentrate on the vertical or internal
relations within the history of philosophy and completely
neglect the horizontal relations (i.e. the ways in which
philosophers are influenced by contemporary circumstances).
Passmore's treatment of retrospective histories is rather
subtle. The point of view of the retrospective historians is
described as: "History is not a mere succession of opinions
which the historian has the responsibility of marking 'true' or
'false', but rather a gradual development towards the truth,
that truth now being fully apparent once and for all" (p.22).
He admits that it is impossible for a historian of philosophy
not to be influenced by the present situation in philosophy:
"Retrospective history .... is the only sort historians
actually write" (p.23). But he rejects as a-historical the
interpretation of the history of philosophy in exclusively
present terms and the interpretation of the present situation
as the all-embracing result of the past.

     Problematic studies are Passmore's favourite ones. As he
sees it, they combine the positive aspects of the other kinds
and are free of their negative aspects. The problematic historian concentrates on problems, not on solutions; and he asks
himself three questions:
a) What problem was the philosopher in question trying to
   solve?
b) How did this problem arise for him?
c) What new methods of tackling it did he use?
By asking these questions the historian of philosophy is dealing with the right subject matter, because "truth has no
history, but the discussion of problems has a history" (p.31).
And he is dealing with it in the right way - establishing both
the horizontal and the vertical relations and emphasising the
permanent importance of a contribution to philosophy. "The
problematic approach .... is the only one which throws light on
the inner development of philosophy; it lets us see how our
understanding of philosophical problems advances" (p.30).

     In Passmore's opinion, however, the problematic approach
is not only the best from the point of view of historiography,
but also because "it is only from problematic history that the
philosopher has anything to learn; it is only that kind of
philosophy which can help him to become a better philosopher"
(p.31).

3. EVALUATION OF PASSMORE'S OPINION

     Passmore's article has at least one undeniable merit: it

provides a reflection upon the historiography of philosophy.
And, unfortunately, reflections of this kind are rare among
historians of philosophy. Moreover, his objections against the
polemical, cultural, doxographical and retrospective studies of
the philosophers from the past are convincing. But we do not
think he is quite right in what he says about the problematic
approach. We agree that, from the point of view of historiography, it may be fruitful to concentrate on philosophical
problems and on the question of how they arose. One then may be
able to establish what the philosopher in question is actually
talking about; and both the horizontal and the vertical
relations can be taken into account. However, when Passmore
says that the problematic historian has to ask what new methods
of tackling the problem the philosopher has used, and again,
that the problematic approach lets us see how our understanding
of philosophical problems advances, we inevitably get the
impression that his pleading for the problematic approach is
based on assumptions which are hardly acceptable as principles
of historiography:
a) that there is progress in the history of philosophy, and
b) that this progress consists in finding new methods of
   tackling problems.
     We do not deny the possibility of establishing progress in
the history of philosophy. But establishing progress demands a
criterion. The criterion suggested by Passmore in his second
assumption contains an idea (his idea?) about what philosophy
ought to be, i.e. concentrated on methods and novelty. It
thereby makes the evaluation of the history of philosophy
philosophical instead of historical: not the historical value
of what happened is established, but the philosophical value.
We do not intend to say that one is never allowed to do so, or
that a philosophical evaluation of philosophers from the past
is useless. But we do not see how this can ever be a good
starting point for historiography. However, there is another
danger connected with thinking in terms of progress: even if
one succeeds in developing a criterion for establishing progress which is acceptable from the point of view of historiography - a very difficult, if not impossible, undertaking in
our opinion - one may very well be tempted to emphasise the
progress in the history of philosophy to the neglect of other
phenomena. The result will be a picture of the history of
philosophy, which may be very encouraging, but at the same time
is too distorted to be qualified as good historiography.
     As a matter of fact, we think that this is exactly what
historians of philosophy have done and are still doing too
frequently - thinking in terms of progress, using their own
ideas about what philosophy ought to be as a criterion in the
evaluation of the history of philosophy and overemphasising the
progress thus established. Jonathan Rée (1978) is right in com-
plaining about this attitude when he writes "Historians of

philosophy project into the past an idea of philosophy as a
professional academic specialism .... they do not know whether
they should be concerned with <u>great</u> philosophy or with <u>influen-
tial</u> philosophy .... and they are so preoccupied with explicit
controversies between philosophers that they fail to notice
areas of agreement or of silence".

4. BIFURCATION AS A MODEL OF DESCRIPTION IN THE HISTORIOGRAPHY
OF PHILOSOPHY

What does bifurcation have to do with all of this? Before
we give an answer to this question we would like to make some
preliminary remarks. Our concept of bifurcation is quite
different from the highly theoretical and well-developed concept of bifurcation in, for instance, mathematics. It shows
more resemblance with the concept of bifurcation which was
introduced in geography by the German geographer A. van
Humboldt as a model of describing the branching of rivers. It
is not our intention, however, to use a trite metaphor and to
present the history of philosophy as a branching river: historiography is not poetry and the history of philosophy is too
complicated a process to be presented in such a way. But at
least one important phenomenon in the history of philosophy may
very well be described by a model which in some respect is
comparable to the model used in geography: we mean the influence exercised by philosophers. If used as a model of description, bifurcation can be defined as follows: Bifurcation is
that phenomenon in the history of philosophy when from one
philosopher various lines of influence go out in various
directions. By the term <u>various</u> we do not only mean <u>more than
one</u> but especially <u>of a different character</u>. Lines of influence
may be strong or weak, positive (i.e. leading to imitation) or
negative (i.e. leading to opposition), and they may stay within
the sphere of philosophy or go outside – to scientists, men of
letters, etc.

But in our opinion it is not only possible to describe the
lines of influence going out from a philosopher with bifurcation as a model; it may also be very useful from the point of
view of historiography. And here the relation comes up between
bifurcation and our criticism of certain elements in Passmore's
problematic approach and in general of the tendency among historians of philosophy to think in terms of progress in philosophy, or in other words: the tendency to use progress in
philosophy as a model of description. We have said before that
this model of description contains at least one serious short-
coming: the picture of the history of philosophy produced by it
easily tends to be too distorted to be qualified as good historiography. If one uses one's own idea about what philosophy
ought to be as a criterion for establishing this progress, the

result becomes even worse. But that is not the main point here.
The quality of progress in philosophy as a model of description
does not depend on the criterion used — the model as such is
wrong as a starting point for historiography. If one compares
bifurcation and progress in philosophy as models to describe
the influence exercised by philosophers one will see why.

Bifurcation as a model of description does not exclude in
advance any line of influence, but progress in philosophy does.
It excludes all the lines going outside the sphere of philosophy and also, within this sphere, all the lines leading to a
result which is not by one criterion or another thought to be
either a step forward or the biggest step forward. The picture
produced by this model not simply tends to be too distorted to
be qualified as good historiography because it is too restricted. Historiography is necessarily based on a present idea of
what is relevant in the past. And in this sense one could say
that every picture produced by historians is restricted or even
distorted. But progress in philosophy as a model of description
determines the nature of this restriction in advance and it
thereby makes historiography inflexible. It deprives the
historian of the opportunity to decide from case to case what
is relevant. In other words: it deprives him of the opportunity
to be a good historian.

The fact that bifurcation as a model of description does
not exclude any line of influence in advance does not imply,
however, that it automatically produces a better picture. But,
unlike "progress in philosophy" as a model of description, it
lets the historian free in restricting his pictures. It therefore gives him the opportunity to use relevancy as the main
criterion. And what is relevant may differ from case to case.
Sometimes strong lines of influence leading to imitation are
relevant and rightly get the most attention. But if all attention is paid to those lines and the other ones are completely
neglected or even rejected, they become relevant and deserve
emphasis.

Let us give some examples. Greek philosophy is frequently
depicted as leading from the so-called physicists or physical
philosophers to the sophists, from the sophists to Socrates,
from Socrates to Plato, from Plato to Aristotle and then
declining to such hellenistic philosophers as the Epicureans
and the Stoics. We do not deny that from a certain point of
view one may draw a line like this; but in our opinion that
point of view has less to do with historiography than with a
judgement about the philosophical value of Greek philosophy. In
this case it may be useful to correct the picture by pointing
out lines of influence which frequently are neglected and
therefore become relevant. For instance, the lines of influence
leading to imitation which went out from the physicists after
the sophists appeared and the other lines of influence which
went out from Socrates beside the line leading to Plato. The

same goes for the picture which distinguishes between medieval and modern philosophy and interprets their relationship as a kind of catastrophe – a picture which is becoming less dominating, but it is still there. In this case too the basis is formed primarily by a judgement about philosophical values. As a kind of counterbalance it may be very useful to point out that, for instance, medieval philosophy is more than a servile imitation of Augustin and Aristotle or that the medieval way of thinking did not stop after Descartes and Francis Bacon.

We are conscious of the fact that bifurcation as a model of description is limited in its use. It only relates to vertical relations and these are only one aspect of the history of philosophy. But within this restricted, though important, field it is in our opinion a means to make historiography more historical. Because historiography may be time-bound and every picture produced by historians may be restricted or distorted in one way or another, this does not mean that these pictures do not have to be historical, i.e. at least not incompatible with what we know about the past. But in the historiography of philosophy the influence of ideas about what ought to constitute philosophy is often so strong that the pictures produced by historians are not as historical as they should be. Bifurcation as a model of description does not suffice to replace these pictures by historical ones. But it is a good means of correcting them and to make them more historical on at least one point – the vertical relations in the history of philosophy.

REFERENCES

Passmore, John: (1965), The idea of philosophy, in: History and Theory, The historiography of the history of philosophy, Suppl. V, pp. 1-32.

Rée, Jonathan: (1978), Philosophy and the history of philosophy, in: Rée, J., Westoby, A. and Ayers, M., Philosophy and its past, pp. 1-39.

SPACE AND ORDER LOOKED AT CRITICALLY. NON-COMPARABILITY AND
PROCEDURAL SUBSTANTIVISM IN HISTORY AND THE SOCIAL SCIENCES[*]

Frank Perlin

Subfaculty of Social History
Erasmus University, Rotterdam

1. INTRODUCTION

It is a matter of curiosity and of considerable significance that two or more centuries after the beginnings of a
powerful movement to introduce irreversibility into European
thought, new attempts should continue to be made to institute
that same epistemological revolution – attempts which are highly controversial and attended with frequent failures and reversals.[1] Nor is the introduction of irreversibility simply confined to the specific difficulties of the natural sciences;
every branch of systematic thought is affected by the same fundamental problem, with the exception of one or two most surprising, but significant cases, to be dealt with in detail in
this essay.
    The beginnings of debate in the social sciences in the
1960's were characterised by a general optimism that changes in
the character of knowledge, together with a general reorgani-
sation of the disciplines, would soon follow the wide-ranging
critiques then appearing. The hopes of those years were not
fulfilled: whether seen in terms of the shape of a discipline,
or of the content of facts and interpretations composing it,
much remains, in all essentials, the same. Instead, the old
interpretations seem to be taking on a new life of their own,
little has changed in the demarcation of one discipline from
another, and a profound scepticism has come increasingly to
take hold of thinking in the early '80s. The change of view
evinced by the late Philip Abrams exemplifies this point; fifteen years ago he could write of an imminent fusion of history
and sociology: their objects of study were the same, and their
purposes and methods were coming closer together; two years

149

M. Hazewinkel et al. (eds.), Bifurcation Analysis, 149–197.
© *1985 by D. Reidel Publishing Company.*

ago, without mentioning this change in emphasis he no longer
believed in the possibility of such a fusion: examples were
listed showing how little real success had been accomplished,
and he doubted whether the loss of the distinct identities of
history and sociology was even desirable.[2].

Now, the point about the social sciences (economics, sociology and anthropology) is that they suffer from the same systematic limitations as has recently been observed of the natural sciences. Anthropology forms a particularly good example,
because the social object chosen as its special field of study
has been seen to possess the same fundamental characteristics
as nature itself – that is to say, the classical Newtonian view
of the natural universe became, in a simplified form, the classical anthropological view of the nature of "simple societies"
(enclosed systems formed by adherence to a finite number of
norms and rules, thus, in essence, holistic and unchanging).[3]
Moreover, "classical anthropology" co-opted the methodologies
and fundamental axioms of the classical sciences as its own
ideal model: Radcliffe Brown's biologism is well-known[4]; what
Lakatos calls the "classical empiricism" of the natural sciences[5] was raised to the level of a general methodological principle by the social scientists; the purpose of the anthropologist was to determine laws of social functioning. isolated from
the random clutter of irrelevant facts in which they were embedded. History was soon characterised as the concern of those
interested in narrative and detail, biography and event; historians were those who at best could do the practical research
which sociologists could later use for their theoretical concerns; at worst, they were embedded in a futile pursuit of fact
for fact's sake, the wood invisible for the trees. Historical
fact was interesting (useful for introductory chapters) but
essentially irrelevant for the tasks of generalisation. Note a
characteristic rule of method: <u>generalisation was not considered incorporative of fact, but selective,</u> a point to which I
shall return later.[6]

A scientistic faith in what was popularly conceived to be
science, could not make of anthropology a properly scientific
discipline; arguably, it has rarely achieved the quality and
profundity of the latter. Moreover, the fact that the classical
viewpoints of the natural sciences have themselves been under
attack since the early decades of this century (I am referring
to the long and complex process of rethinking that has taken
place since relativity and quantum mechanics began to be elaborated) should help the reader to recognise that the problem of
irreversibility concerns human thought in all its aspects, and
further implicates the problem of investigating and characterising its underlying unity. Each of the systematic sciences defines its universe (the universe implicated by its conventional
body of methods of approach and methods of making statements,
despite all particular qualifications made by its practi-

tioners) as fully formed in so far as its fundamental features,
worthy of analysis and characterisation, are concerned: the
laws that characterise them are predicative of the states of
being in the future, which indeed form an unchanging continuum
with past and present.[7] These laws are "facts" of a higher
order than facts of a merely phenomenal kind, and indeed both
may take on the character - because the latter are indisputably
part of every day experience - of a contrast between the real
and the apparent. Of course, there are many ways in which, even
within one discipline, this contrast is presented, but all possess a fundamentally dualistic vision of the universe (even if
not consciously formulated as such). Einstein's position on
change, cited by Prigogine, is symptomatic of epistemological
problems that have survived all "revolution" in the development
of scientific thought.[8] We can sum it up by noting that in
spite of the well-known classical confrontation between Newton
and Leibniz over the structure of the universe and the nature
of knowledge, the "neo-Newtonian" conception of the natural
universe inherited by the sciences (social and natural) continues to possess an essentially Leibnizian character, that is
to say a Platonistic dualism between the essential and phenomenal.[9] The "laws" of the social sciences too, laws formed
out of a still popular image of natural science, which its own
philosophers have long since rejected, seem to possess an existence discrete from that of phenomenal matter.[10]

Now, it seems to me that the very problems that have long
characterised the conduct of history, and concerning which colleagues in other disciplines have traditionally waxed considerably contemptuous, turn out to be its lasting strengths (whereas the apparent strengths of the social sciences - their concern for the essential and lawful, have become increasingly
controversial and the subject of intense debate). Mainstream
history has long been characterised by its fetishistic attention to detail and chronology (for the moment, in representing
the case for history, I shall accept the "worst-case" profile
of the discipline, as projected by social scientists). In constructing a biography, a course of political events, or the
rule of a particular king, and even in its more analytic forms,
we have been burdened with the most naive conceptions of relationships between the cause and effect, or of how events take
place. The very lack of theoretical sophistication - or better,
a characteristic contempt for theory and philosophy - has permitted the general body of traditional historians to concern
themselves with the study and presentation of detail for detail's sake.[11] Given the ethos of history as the time-bound
reconstruction of past happenings, the consequence has been a
concern for careful reconstructions of finite, observable
changes of manifold kinds. One should also emphasise that the
significant unit of historical knowledge is not so much the
individual study (the scholar) as a body of studies (thus a

population of scholars) working around a particular problem. It
is this corpus of disagreement and precisions over exact sequences of particular courses of events and their detail, that
renders change as the inherent focus of the discipline (its
"discourse" so to speak).

Of course, we no longer agree with the old explanations,
which could often be of very low quality, but the point is that
due to this extreme empiricism, we have inherited a body of
knowledge, a set of methodologies, a conception of generalisation, that form a remarkable exception to the other disciplines: in history – but we need to qualify it as the history
of Europe – irreversibility lies at the root of our method, so
that, by contrast, static conceptions of societies tend to be
the subject of special, extremely controversial theories (frequently the application of models derived from the social sciences or from a natural science).[12] I am implying that, given
certain assumptions, the naive empiricist ethos of historical
writing fostered an approach that departed radically from the
methodologies of the more systematic and positivist social
sciences, constructing a kind of knowledge in which irreversibility was inherent.

It needs to be added that this image of history and historical scholarship thrown up by the social sciences, is a parody
and necessarily so. This parody of the discipline is identical
in kind with that of the diachronic itself (the "phenomenal",
change, individual and personal acts and interests, events),
which itself formed a necessary aspect of the idealisation and
isolation of what was deemed to be the essential subject matter
of the social scientists, the synchronic (functions, structure,
laws).[13] A fair proportion of historical writing has indeed
possessed this character (and not only because historical
writing has always included a large number of amateur enthu-
siasts), but what was completely ignored was a significant
amount of writing that combined together an empirical approach
geared to observing change, on the one hand, and a concern to
answer important theoretical questions, on the other. One may
mention Vinogradoff, Kosminsky, Postan, Bloch, Seebohm, Gray
and Unwin, among many others writing in the early decades of
the 20th century at the very time when structural functionalism
was taking form, and whose works and views remain of formative
significance for the branches of history concerned (although we
no longer always agree with the explanations and general models
of society contained in them). This apart, it remains true that
an interest in theory is not the same as a capacity for self-
reflection concerning the nature and methods of the discipline
itself. The absence of the latter has meant that historical
knowledge has largely depended upon an essentially unnuanced
characterisation of history as the study of succeeding events
in past time, and correspondingly of an empiricist attachment
to the study of "microscopic" events and relationships of a

fairly extreme kind. Under such conditions (the isolation of
history from the general developments affecting the social
sciences), empiricism could possess an unexpected value.[14]

From this perspective, my judgement of Prigogine's "theory
of bifurcation" is ambiguous, and ultimately negative. I am at
one with much that he argues on the philosophical level: for
example, the unification of knowledge, the move from a world of
quantificationism to one of structuralist pattern formation,
the view that non-equilibrium makes a system extremely sensitive to external conditions (matter incorporating all the dis-
symmetries of the outside world), the shift from a conception
of the future in terms of trajectories to one at best characterised by probabilities, these form part of a broad movement
of really fundamental change that has begun to gnaw at the fab-
ric of our modus operandi.[15] Marx's history of the rise of
capitalism in 19th century England, and certain aspects of the work
of Levi-Strauss and Foucault, are notable examples, but these
names – their controversial character, the fact that for mainstream history and social science they tend to remain peripheral and parodied – indicate how difficult this transition
is; one that certainly will not succeed by its own momentum.

The difficulty comes in the translation of this philosophy
to the realm of method and hypothesis, and of concrete statements about the practical relevance of bifurcation for the different disciplines (in the case of the examples treated in this
volume, for physics, geography and economics). To my mind,
these statements raise more difficulties than they solve. A
viewpoint which states that we should move from the classical
perception of the universe, formulated from an epistemology
that has excluded change, to a new perception based upon non-
entropy, irreversibility, non-equilibrium, is absolutely different in kind from one which states that under certain conditions, physical systems change from states of equilibrium to
non-equilibrium, that the direction of change has a random
element in it, and that the change itself involves transformations that are unpredictable (all of which define bifurcation).
The former states that the universe is not "this" but "that",
the latter that under given conditions we may observe shifts
from "this" to "that", that is to say between states. Prigogine's philosophy is fully in line with the philosophy of the
Romantic movement: take the remarkable similarity between his
elegant view of time, as not so much an externally derived
linear measure, as part of the very character of changing matter (the chrono-topography, as he calls it, of an urban complex
as developed in the course of time), and Oken's (1810) "Time is
nothing but the Absolute itself", and "The Absolute is not in
time, nor before time, but is time".[16] As Lovejoy puts it,
"The realization of God, then, takes place only gradually
through the history of the cosmos. Its primary manifestation
and universal condition is time".[17] Here is an example of the

displacement of the duality between essence and phenomena
through the many-levelled process of unification of knowledge,
the reduction of those inconsistencies and contradictions which
Prigogine rightly rejects.

Bifurcation, however, is different. To the extent that it
forms an ordered concept (I intend to criticise this aspect of
it below), it seems to me to stem from the very specific problematics of the model building sciences, in which states of
stasis and equilibrium are fundamental to the manner in which
their different contents of thought have developed. We have to
ask ourselves to what extent we can give to changelessness the
status of any real existence, or, more to the point, the extent
to which even a conventional commitment to methodological
changelessness is reasonable. We can assume that it exists or
that it is reasonable, or that it does not exist or is not
reasonable, but either way the assumption leads to inescapable
consequences. If one takes the stand that in a large variety of
cases, and potentially all, changelessness is not discovered
but introduced, often without discussion, as a formative principle, the methodological consequence should be that no theory
of knowledge can take it for granted, even for convenience,
without expecting pathological results (or at least without
showing serious concern for its possible consequences).

What I shall try to do below is to demonstrate that bifurcation stems from an epistemology grounded in the classical
view of the universe. It is one of a number of supposed solutions to the problem of irreversibility, that, in fact, fails
to grapple with the essential problem, because still retaining,
within its core, central, axiomatic aspects of the old knowledge.[18] In my view, bifurcation may be not merely irrelevant
to the problems it seeks to solve, but, in the case of history,
would involve a notable step backwards from its current position.[19]

An analogous difficulty occurs in anthropology, the problem being the question of what sort of status critics should
give to anthropological knowledge itself. Marxist anthropologists (worth mentioning because of the ostensible Marxist
commitment to dynamic models of society, with which I am entirely sympathetic) were at the forefront of the first wave of
criticism concerning the nature of the discipline, and of its
characteristic interpretations of simple societies (a wave now
considerably broadened). I mentioned anthropology's naive form
of scientism, that set its purposes as the discovery of "laws"
of human organisation. The corollary of this programatic attitude has been the actual empirical discovery of societies un-
affected by historical change, characterised by a number of
symptomatic "general" laws of social organisation: reciprocal
exchanges, kinship determinism, ideological determinism, and so
forth. Despite being hedged around with qualifications and even
denials, the kind of thinking concerned provides no room for a

departure of behaviour and conscious dispositions from either
institutionally-determined or culturally-determined norms,
rules and moral codes, with the result that each such "society", however differentiated, takes on the character of a
virtual equilibrium condition, a state of being that by its
logical nature excludes an intrinsic meaning for both time and
context.[20] Since the discipline of history and historical
facts are both considered essentially irrelevant for purposes
of generalisation, the broad historicist assumptions implicit
in such dogmas have never been tested (absence of change being
a statement about history: a case of essentially historical
statements in a historical vacuum shaped by ahistorical dispositions).[21] Naturally, this is a simplification of a highly
complex topic, but an avid reader of anthropological and historical works will be well aware of how deep and profound such
views are, how ubiquitously they appear in often brief but
always significant asides, indicating the basic axioms which
colour interpretations and methodology alike.

Now, to return to the critics of the discipline, a number
of anthropologists (who, although critics, were, like the bifurcationists, trained within the terms of the disciplines they
have come to criticise) have attacked these dispositions, but
at the same time hold on to the classical knowledge that has
come to be associated with the latter (the fundamental axioms
that in effect seem to justify the separate identity of anthropology as a discipline, which may also be summed up as essentially the different laws of social organisation and passage-
through-time seen to distinguish non-Western from Western societies). In this respect, their criticism is merely "super-
structural", failing to consider the essential connections
between the shapes and extents of an accumulated knowledge, on
the one hand, and the methodologies and theories they dislike,
on the other. Their criticism rejects the culturalist structure
of thought characterising anthropology in favour of a more ma-
terialist conception of society, but accepts the given empirical content of the discipline.[22] The results are new interpretations which do little but vary old themes and solve few or
no problems (Godelier's dictum that kinship and culture form
part of the forces of production being an elegant example).[23]
In most cases, new interpretations have failed to lead to
fundamental change, because they have failed to stimulate a
revolution in the attitude to what should constitute a suitable
stock of knowledge and of approaches capable of testing the
basic hypotheses of the discipline. Herein, also, lies my ob-
jection to bifurcation, which also, without reconstruction,
incorporates central aspects of the classical view of the
universe as part of its own perception.

The problems becomes clearer when anthropology is used to
illuminate European history; then, the static, totalitarian
assumptions and methods of the anthropologists come face to

face with the dynamic conflict-oriented approach of European
history writing. Gilbert assesses one such case, but makes a
fundamental error of a kind similar to that noted above, in his
rightly critical reaction to an anthropological treatment of
Renaissance Florence, an interpretation in which social conflict, cultural differentiation, event and chronology have no
essential place.[24] He states that its author mistakenly applies models suitable for non-Western societies, such as Bali
(in this case Geertz's "theatre state"), but unsuitable for the
"Western world with its peculiar political and economic dynamism". But, this confrontation with what we know about Florence
may in fact constitute a basis for a more thorough-going critical examination of anthropological thought as such, and not
only for those societies where sufficient historical research
exists to disprove it in a particular case. In this respect,
the major difference between Europe and the third world is
that, due to the characteristics of historical scholarship outlined above, Europe is a rare case where sufficient evidence
exists to contradict the basic axioms of anthropological
thought, and to attack those assumptions upon which its traditionally ahistorical approaches are based. It is notable that
the anthropological vision concerns a residual category of both
complex and so-called "simple" societies, essentially defined
by their being non-Western. It is the discipline that has discovered the knowledge that pools what are, at this level, essentially incomparable kinds of societal organisations, a
knowledge not found where different methodologies and assumptions have been at work.[25]

      Moreover, recent historical research, impelled by new
types of question, have begun to provide an alternative body of
knowledge contradicting the old dogmas about India and Africa,
locating societies subject to complex organisational changes
long before colonial times.[26] This does not mean that historical research is <u>necessarily</u> superior to that of the anthropologist, but that <u>it is demonstrably</u> so in these cases, because
its surprising findings are based on research of problems
neglected by the anthropologist and thus encompass a broader
and more detailed knowledge of the questions at issue, capable
of testing anthropological precepts.[27] The Western world is
not so much peculiar in its dynamism, therefore, but in having
been the subject of the peculiar discipline of history. Now,
the purpose of arguing this case lies in its exemplifying the
difference between a critique of the validity of a type of
knowledge, and one that merely displaces that knowledge to a
restricted part of its universe. The point is that of showing
the need for a shift from Prigogine's technical formulation of
bifurcation, where both states of being appear to possess a
real existence, to the level of his philosophical critique,
where the concept of irreversibility is seen to have inescapable consequences for all knowledge.

In the following pages, I shall explore certain possible
applications of "bifurcation" to historical explanation in
order to illustrate these objections. Then, I shall make a
short critique of approaches to space and order used by sociologists and, to some extent, historians. The resulting text
should enable us to conclude that equilibrium states and random
states in human societies derive from a faulty epistemology.
Since it is Prigogine's philosophical departure that interests
me, and not his theoretical solution (bifurcation), I intend to
place of strong emphasis on the former as the true (if largely
hidden) concern of our volume.

2. SOME WEAKNESSES IN THE PRESENTATION OF BIFURCATION, AND THE "DISCIPLINE OF CONTEXT"

Besides the conceptual problems raised above, to be
further elaborated in section 4, the presentation of bifurcation by Prigogine, Deneuborde and Allen, seems to me unconvincing, and, in a non-pejorative sense, essentially incoher-
ent.[28] One needs to be already convinced that one is on to a
good thing, in order to see these different, essentially incomparable situations as confirming a fundamentally new departure,
bifurcation.[29] The following argument will attempt to justify
this point of view.

Prigogine rightly remarks on the fundamental importance of
micro-studies for providing unexpected results:[30] this observation can also be applied across the board to history and anthropology. In each case where the classical models of reality
are at their most dogmatic, there is a significant lack of precise knowledge and a corresponding insertion of what Bourdieu
rightly calls "special solutions" (or special logics).[31] It is
here, especially, that micro-studies have a particularly important role to play in helping to redefine the basic perceptions
characterising a given discipline. But some qualification is
also necessary. In the case of India, for example, the anthropologist's substantivist model of economic exchanges has co-
existed with a lack of research on the questions of money and
markets.[32] Meanwhile, the historians have been slowly building
up a small body of new knowledge which contradicts the substantivist position on these subjects. Yet, the divide between the
disciplines, between cultural substantivists and economic historians, far from diminishing in this period of interdisciplinary aspirations, is being increased by this access to new
knowledge. The apparent lack of interest, among anthropologists, in the results of new research of this type is turning
out to be a more pathological problem than suspected: there
seems to be a tendency to reject inconvenient findings and
"topics"; the old models seem to be so central to the systems
of thought concerned that the results of micro-studies appear

irrelevant, a situation apparently paralleled in experimental physics.[33]

The legacy of these contradictions has also created a curious state of ambiguity among historians of the third world, who in radical contrast to the Europeanists have been deeply influenced by anthropological ideas. Correspondingly, they have tended to lack the confidence needed to translate their critical views into concrete research strategies (because their criticisms tend to be provisional and undermined by general assumptions). For example, there are those who claim that when taxes are demanded in the form of cash, peasants become integrated into a broader nexus of market exchanges; others state that once the taxed portion of the agricultural product is exchanged for cash and paid over to the tax collector (presumed to be a question of events on one or two days of the year), the peasant returns to his closed, autarchic, moral economy.[34] Neither view is correct, but what is important is that the solution to this problem is surely simple: the conducting of research into this question, since sufficient documentation indeed exists.[35] This is one of a large number of examples of areas of ignorance, the existence of which turns out to be of major strategic value for the survival of the otherwise discredited assumptions and approaches of a discipline.

What I am pointing to, therefore, is the tautological relationship between the various components of thought, method and research in the different disciplines: "micro-studies with surprising results" depend on a framework of thinking already disposed towards observing and taking seriously such results. If Feyerabend is correct concerning the inadequacy of the lenses used in Galileo's telescope, quite clearly there can be an irrational element even in the "surprising results" of the micro-study, but given the novel cosmological framework of thought with which Galileo was operating, that element is less irrational when viewed in terms of the predictive power of improved syntheses of previously contradictory information (at least, its irrationality only exists in terms of the logic of the micro-study itself: a mystique of method which creates a false genealogy for these "results").[36]

"Bifurcation" has similar characteristics in that, together with the micro-studies which form part of it, it stems from new philosophical dispositions towards matter, even though its actual experimental status may remain controversial. The problem arises if there are too many inconsistencies of the kind argued above. It may be emphasised that the Galilean "revolution" ultimately resulted from the nexus of new precisions involved in a process of what we may call "contextualisation".[37] A great deal of the fundamental critique now opposing anthropology's cultural substantivism derives from a similar incorporation of precisely worked out contexts into the field of relevance. Does bifurcation succeed in following through such a programme of "contextualisation" to completion?

Here, some comments on the experiments treated at the conference are in order. Prigogine offered a jar into which a mixture of liquids had been introduced, and asked his audience to observe the various patterns which successively arose in it, to note the structuration of the content of the jar, and to conclude that there was a random element (unexpected and unpredictable) involved in the process of mixing. But the appropriate response should surely have been that we were not provided with sufficient information to decide on the issue concerned, just as the participants at Galileo's demonstration lacked sufficient information (or false information, stemming from the mystique of the method referred to above), to make an independent judgement of the validity of his claim.

Bifurcation, a critical point at which a "choice" in direction of change occurs, is apparently exemplified by the transition affecting amoeba when subjected to deterioration of food supply (transition from separated organisms to a collective body of amoebae); by the transition among termites from random individual building activities to co-operative construction (the construction of a nest); by the choice taken by ants to use one of two exits in an apparently random situation.[38] The case of the amoeba typifies "bifurcation" as a point of transition largely resulting from changes in context; the case of the termite involves a critical point reached through collective accumulation of actions in a relatively limited space; that of the ants, however, seems to require us to consider "choice" on the most superficial of levels (one arising from the chance accumulation of larger deposits of chemical substance at one exit in the period of random selections), in Hobsbawm's words, "Jewish jokes in which every situation contains two possibilities".[39] Apart from the weakness of the last example (in which we can neither talk of a change of state nor of a response to mutations in the context), it seems to be unclear why we need to introduce the notion of bifurcation in order to understand a transition to collective dispositions, or a change in scale of the unit of activities concerned. These seem to be situations which are recognisably determined (or at least possess the potential of being so); "bifurcation" essentially describes the character of the scientist's observation before he has achieved an explanation: he observes a change that has no apparent necessity inherent in the nature of the constituents; he examines the contextual disposition of relationships, and discovers the deterministic links between them and the precise conditions in which the experiment occurs. Then it is time to forget about bifurcation in this merely technical sense.

The historian has an identical problem, well described by Hobsbawm in his essay on predictability.[40] Outside of the old historicist dogmas of a Toynbee or a Morgan, the "discipline"

of history has always seemed to be typified by the random
character of the changes involved in the histories studied.
There has seemed to be an unbridgeable gap between concrete
event and individual action, on the one hand, and generalisation, on the other, with the result that historians tend to
argue that "proof" is never possible: we accumulate probabilities (or means of disproof) rather in the way that Popper
argues. However, one of the major changes occurring in the conduct of historical explanation concerns the degree to which
these problems are becoming understood, the observation of what
in our (already formed) practice is significant for deriving
explanations. Given our lack of laboratory-type situations, our
principal problem is coming to be seen as one of seeing order
(but not stasis) in the irredeemably complex web of an open
frontier of events, actions and relationships. And it is the
emphasis on the last element — the open frontier, or science of
context, in radical contrast to common sensical tendencies in
virtually all branches of "science" towards enclosing the
framework of apparent relevance — that is beginning to alter
the manner in which we understand history itself, thus what is
really involved in "theory" and "generalization" about society
(with a consequent shift from a selective notion of generalization to one that is incorporative). What this implies is a
shift from a situation in which "bifurcation", in a figurative
sense, stands for a characteristic mode of explanation, to one
where the essential means for providing an explanation are
better and more precisely understood.

What we need to emphasise is that the extent of this complexity, taken together with the inadequacy of our means for
dealing with it, makes the predictive power of history very
poor, a poverty deriving from the limited extents of both theoretical and empirical knowledge; "bifurcations" (choices of
direction) reside in our own limited means, but not necessarily
in the object itself. The Russian Revolution did not arise from
the chance circumstance of Lenin's return to St. Petersburg,
nor from his decision to seize power at a certain moment (although both these alternatives have been argued, and although
these affected the actual detail of what occurred). Far from a
destabilisation at a point of rupture historians, in fact,
argue over the question of when, in the 19th and early 20th
century, social conflict and instability in Russian society
become sufficiently significant to become a necessary component
of a convincing hypothesis. G.T. Robinson's classic account
begins in the 16th century, and not without reason, while
Trotsky lays much more emphasis on the long term destabilising
effects of the developing international context.[41]

This problem of explanation is crucial to the following
argument, and I shall thus take it a step further. A pioneering
philosophical departure, in this respect, was that of Sartre's
Critique of Dialectical Reason, in which he elaborated a theory

of the contradictory, but interdependent relationships between
individual actions and purposes, and the structural characteristics of societies that define those individual ends and yet
subsequently defeat them through unsuspected changes collec-
tively resulting from those same actions; thus the manner in
which these opposite scales of social "event" (particular and
general, agency and structure) inform one another in unpredictable but deterministic ways.[42) This, in my view, is a successful identification of the fundamental problem for any effective
theoretical treatment of the problem of irreversibility and
process in history, quite apart from the question of whether
Sartre succeeded in working it out in practice.[43)

Bourdieu has introduced a similar concept into anthropology with his notion of the habitus, and his attempt to temporal-
ise anthropological method and observation.[44) That these contributions to explanatory principle are not simply passing and
peripheral events, with few or no practical implications, can
be seen when we examine an area of empirical historical research, which has classically been the subject of the types of
"special" explanation which are ordinarily characteristic of
anthropology (at the core of anthropological thought): the
study of inheritence systems and their relationship to demographic process. The first was subject to the view that custom
and law were the determining factors; the second subject to the
Malthusian (and now neo-Malthusian) dogma that explains population growth by the natural, therefore irrational propensity for
individual families to multiply to the maximum possible extent,
so that mortality is the chief variable in population movements. However, recent research has shown that individual inheritance practice, law and custom may all be far apart, and
that the first may be decisively influenced by social and economic contexts, that is to say by rational considerations affecting conscious decision-making.[45) It is precisely in this
area of rational choice that a non-Malthusian demography is
beginning to be written, and the emphasis of explanation to be
placed on fertility instead of mortality: age of marriage,
duration of lactation, size of families, inheritence practice
(thus how many of a given family remain within the community).[46) The result is probably one of the most fundamental
departures in historical method (if not yet observed in this
light: this is the accumulation of a concrete body of historical studies, with the usual mistakes, false premises and du-
bious interpretation, but which successfully synthesises the
fundamental link between levels of observable historical change
(although this is a process which is by no means complete, of
course). Nor is this to impose a false, psychological (and
equally substantive and irrational) rationalism of a Hobbsian
kind[47): the explanations for individual dispositions derive
from micro-studies of community and context, that is to say, on
the one hand, of information concerning social and economic

conditions (thus, significantly, involving a breakdown of the
boundaries traditionally surrounding historical demography),
together, on the other hand, with a study of collectivities in
which individuals can be isolated for observation.[48]

This brings us back to the question of the micro-study. My
remarks above were not meant to imply that the micro-study did
not have a crucial role to play in effectively altering our
knowledge. Historical attitudes change, but if methods remain
the same, and new studies are not undertaken they are surely
superficial and reversible. Ten or more years ago Indianists
were speaking of the myth of the enclosed, self-sufficient village; there seemed to be a concensus on the falsity of this
long-standing dogma. For the most part, however, in the absence
of a corresponding revolution in method, work after work im-
plicitly reformulates the closure of the Indian village, so that
there seems to be a tendency to return to the old dispositions.[49]

Philosophically speaking, we have described the development of concepts of history that not only assume irreversibility, but are now beginning to be able to deal with that notion
in more precise theoretical terms. Sartre, in effect, was describing the inherent transitionality of all society.[50] Notions
of equilibrium seem to be the victim of these changes. Yet, in
the present transitional phase it is clearly tempting to draw
relationships between the two states. For example, some historians still argue that demographic equilibrium broke down in
the face of new occupational opportunities in the countryside
in the 18th and 19th century, giving rise to the new processes
of demographic growth and behaviour of the Industrial Revolution; here, a change from Malthusian to non-Malthusian theory
is reified in terms of a change from pre-industrial Malthusian
conditions to the non-Malthusian conditions of industrialis-
ation (until, perhaps, a new equilibrium is reached).[51] This
is much in the same way, as we argued above, that Prigogine's
philosophy of irreversibility is reified in terms of bifurcation.[52] Instead, what we prefer to argue, in line with his
philosophical disposition towards irreversibility, is that the
experiments described by the bifurcationists treat transitions
between one state of change (or set of structures in which
change occurs) and another. This, indeed, is how we would describe the effects of the Russian Revolution. It is the laboratory situation, with its conventional start with discrete,
and given individuals, that is predisposed to falsify the
meaning of what is observed (from discreteness to collectivity,
autonomy to penetration and integration).

3. BIFURCATION AND HISTORY

3.1. Bifurcation 1, or counterfactualism: a bad case

A simple and straightforward application of bifurcation to

history would surely lie in counterfactualism, which developed
in its more rigorous forms, after the second world war, as part
of a transition of economic history from a branch of history to
one increasingly subject to the methods and questions of economics (with a corresponding application of economic thought, and
a reliance upon sophisticated quantitative and methodological
techniques).[53] In fact, a large number of historians have
failed to accept the validity of the basic assumptions underlying its use, while many others simply ignore it, whether in
its essentially descriptive and polemical form, or in its more
"rigorous" methodological form. The reason can be explained as
follows. Counterfactualism - either, in its "weak" form, the
idea that a particular historical development might have
followed a different course, given the absence of a particular
factor (would the democratic liberal regime have stabilised in
Russia if the Germans had not transported Lenin back to St.
Petersburg?), or, in its methodological form, that the actual
historical development of a society can be illuminated by examining the consequences of the absence of certain critical features - a kind of simulation of laboratory procedures (thus,
Fogel's question of what would have happened to the U.S. economy if railways had not been built) - is based on a patently
inadequate, even false view of historical process. Counterfactualism depends on an over-simplified concept of cause and
effect, and is surely the typical product of methodologists
whose disinterest in macro-theory (at the level of social
structure and evolution, or of that of clarifying the necessary
grounds for formulating particular approaches and arguments),
renders them liable to accord the historical process with an
arbitrariness which closer examination would show to be impro-
bable, but which, nevertheless, seriously debilitates the quality of interpretation and understanding in a given case.

　　However, it is now increasingly understood that no historical process can be understood without an examination of its
general contexts of relationships, and especially as the consequence of a structural development of those contexts in the
past - what Thompson calls the discipline of context.[54] I
shall try to explain the reason for being dogmatic about the
validity of this point, below; for the moment note that it
accords with Prigogine's view of matter incorporating all the
dyssemmetries of the outside world, and with Dzerdjeevski's
climatological argument, that all levels of phenomena, from the
smallest fluctuations in daily weather to the cosmological, are
ultimately connected (in an incorporative and causal manner).[55]

　　The implication of this statement is that the origins of a
revolution or of deindustrialisation (our two arbitrarily selected examples) cannot be explained without the synthesis of a
host of other connections and developments; it is interdependent, even determined by its temporal and structural contexts
(which is not to say predictable).[56] This would also be the

case for small scale events and individual acts and dispositions; even given a range of choice, and thus the possibility
that a certain course of events might will have gone differently, these have to be assessed at the level at which they occur,
while the actual range of different possibilities, and the extent to which micro-events may have decisive influences, can
only be understood in terms of the contextual relationships
linking the different scales (individual and society, event and
a longer historical perspective, situation and a larger geogra-
phical perspective). Changes and reversals may occur during the
course of development, but there is no "point of instability"
at which a major course of development may suddenly be inaugur-
ated, or at which a single arbitrary factor can be implicated
as the cause of a major development (Lenin in his sealed compartment).

It is the banality of the assumptions on which this example of possible bifurcation is based (in both its popular and
more rigorous forms) that leads me to view counterfactualism as
a "bad" and essentially uninteresting case, despite the fact
that it apparently fulfils somes of the conditions required by
bifurcation. A certain range of possible examples can be excluded from consideration, because they constitute either false
problematics, or no problem worthy of novel nomenclature at
all. Deneuborde's experiment with the ants seems to fall into
this latter class, counterfactualism, in both its forms, into
the first.

3.2. Bifurcation 2 and segmentary thought: a better case

A more convincing case of bifurcation than counterfactualism would be the shifts in scale, reference system, personal
identity and kind of ideology, involved in a complex segmentary
system of social and administrative praxis. I am referring to a
cultural complex characterising certain pre-industrial societies, such as had existed in pre-colonial Maharashtra (in Western India).[57] In such a system of segmentary thought and action a large series of classifications - of property types,
administrative categories of taxation and space, settlement
units, different contemporary ways of differentiating the same
population clusters (by caste, lordship, and the like), for
example - intersected one another in a manner which permitted
contemporaries to shift level, subject, criterion, and pattern
of thought, at will (and by choice), without their having formulated, or having access to, an overall theoretical synthesis
capable of explaining the relationships between the "systems".

Logically speaking (on the level of particular, finite
categories, and not the systems of reference themselves), it
may be argued that the definition of the individual categories
and their separation from one another, occurs prior to their
investure with an ideological content. This is simply to say

that once formed, a category (say a village) was likely to assume a life of its own in connection with the complex of relationships in which, at any moment, it happened to be embedded. I term this system "substantivist", a kind of contemporary positivism, but one realised in distinct institutions as well as being conceptualised in particular cognitive forms - a system of praxis, in short.[58)]

For example, old order property was highly heterogenous, consisting of numerous sub-types (land, office, wealth in goods or cash, each of which was combined in different ways, and given names possessing strong social and ideological resonance), which themselves were comprised of many, often very many separately defined and named constituents. Village headmanship, for example, was hereditary and saleable, composed of lands of different kinds, moneys and levies in grain, and office. In order to reduce headmanship to a monetary value within the larger right-holding complex and administrative organisation of the countryside, each constituent had to be specific and computable (that is to say, defined and measurable). Property thus has many subdivisions and types of constituent.

In addition, apart from this type of "real" property, which, being hereditary, was the source of egalitarian, individualist, social and political ideologies, were king-centred systems of rights which were neither hereditary nor saleable, and which formed part of ideologies of lordship and subjection.[59)] These two different property orders were instituted within the same villages, sometimes upon the wealth of the same fields. Note that such rights (of either system) could be incident directly in the field, in a share of the income derived from it, or even more indirectly in a share of the income of the village, region or state, thus at different levels of "synthesis" in the vertical organisation of society.

I am attempting to describe a multidimensional complex of types of intersecting classification which provided remarkable flexibility to contemporaries. An administrator could shift between types of property, or systems of right, and from the level of a field to that of a region, or, with the aid of monetary values, pull out properties scattered in widely separated regions and reduce them to a single numerical symbol.

Secondly, any one individual might possess properties of different kinds and scales, scattered in different districts, or possess different properties in the same location. A particular person may be equal to another (in an unqualified, ideological sense, expressed in actual institutions) through holding one type of property, and subordinate to that same person through the holding of a different kind of right. As is also well-known in pre-industrial property systems in Europe, it is property (that is, the ideological complex subtending on property) that gives a man a <u>social persona,</u> his individuality, so

that if he holds two properties he has two intersecting persona.[60] A further and fundamental complexification was that
there were different ways of perceiving differences between
persons. Caste and religion divides, but persons of different
caste and religion might have identical kinds of property and
thus be considered equals (property brothers), with a code of
behaviour to match.[61] And I could continue.

The point I wish to make is that shifts in criteria involved different forms of identity in a system characterised by
a variable set of individual and collective identities. A shift
in the scale of the criterion used could involve a change in
the order of information required.

Let us also look at these systems in terms of the individual. A property owner is an individual owner; ownership provides him with the occasion to realise his inviduality (think
of the European craftsman allowed to practice his mystery because he is a member of a guild), and makes him part of several
collective individuals (think of the guild itself with its uni-
fying symbol, the chaper and a patron saint). A property owning
person in Maharshtra in the 17th and 18th century is one who
owns a property individually and, by virtue of owning that
single property right, is simultaneously one of a class of
owners of that type of property, or of all the property owners
of a particular village (which had a name, and the grouping a
legal term), or a region (which meet in judicial assemblies and
also has a term), or of property owners in general (the sharers
of the habitational territory, "settled" by property-related
institutions). Or he has more than one type of property, and
thus has two or more individual identities. Or he belongs to
two contradictory but coeval systems of right: the egalitarian
system of real properties, and the other which is hierarchic
and descending, and which is not a real ownership right at all,
but certainly a system of social and political identity contradicting the egalitarian system. Or he possesses several social
identities, a member of this or that caste, a possessor of property, etc. - systems of identity which, to repeat, cross-cut
one another without synthesis. In substantivist systems, in
general, no synthetic mode exists which would enable contemporaries to explain and unify the differences - rather these
function together without the need for that synthesis.

A specific category is called forth by a particular conca-
tenation of contextual conditions - in fact, several categories
may be brought into active play. The government writes to the
headman of a village; this headman happens to be a brahmin, a
government minister, and a powerful military chief. However,
the letter is addressed to the headman of village so-and-so,
the other facts (identities of the person) being irrelevant to
the task in hand. The king is the source of descending power,
but as a possessor of egalitarian properties he attends a judicial assembly of equals (brothers), and is addressed by his
property-vesting title.

The system is coherent, flexible and always in motion.
Being always in use, new categories are invented and others
modified (new properties are created, others altered in content
and structure), while the overall character of the system
evolves in the course of the 17th and 18th centuries. In no
sense do we have a stable system, an organisation of reciprocities or equivalences; being praxis, it can only be understood in terms of its social use and its sociology, not (as the
structuralists might prefer it) in its own isolable terms as an
autonomous self-explicative "structure".[62]

The "memory" of the system (to use Prigogine's metaphor)
consists of this multi-ideologised body of practical knowledge
- its actors with their particular skills and know-how, their
ability to feed in a sign and pull out certain sets of information, and the overall stock and arrangement of classified
knowledge itself, its administrative and linguistic forms, its
institutionalised forms. Variation in situation might produce
different choices of the same scale or order (Deneuborde's ant
paths; a grouping of either brahmins or headmen), or a different level of organisation called upon by changes in scale
(termite nests; the regional assemblies of egalitarian property
owners with their "dangerous" political overtones for the hier-
archs of the royal courts). It would not be misleading to term
this system a kind of primitive computer, consisting as it does
of a huge and complex memory bank of variously classified information.

But it is as wrong to call this random as it is to see it
as part of an enclosed and stable system. It is fixed in recog-
nisable and purposeful institutions and activities, which are
themselves the medium through which change occurs. Individuals
are not autonomous beings whose actions are purely arbitrary,
but are themselves reflective, synthesising and creative points
of a given nexus of structural conditions, so that although
possessing choices in what they do, the nature and range of
choice requires, and is, susceptible to analysis. The individuals were masters of their system in these qualified senses
(operating but not, of course, directing it; "microscopically"
causing micro-change, though not able to predict outcomes; subject operators but social objects of the societal complex in a
state-of-becoming). Change can be studied in the given documentation of events, so that courses of action, and institutional usage and mutation, may be reconstituted (say, a lord
manipulating egalitarian property ideology in order to gain
more access to wealth and power); individual actions involved
in particular events, separated in time and space, and causing
small scale micrological changes, can be seen in sum to compose
a more or less broader process of change affecting society at
large and moving in the same general directions (the "in sum"
and the "more or less", of course, being fundamental qualifications): a developmental or irreversible movement in the
structure and nature of the system, therefore. Micro and macro

are, in such a formulation, profoundly linked, although the
connections are indeed difficult to reconstruct and operate at
and through many levels. <u>This unity is what, in my view, makes
any given social system inherently transitional</u> (the unity
between acting, choice-making subjects and structured contexts,
with its systematic generative contradictions). By contrast,
given the ahistorical assumptions held by anthropologists about
the "societies" they study, it is understandable why the
choice-making subject could possess no essential formative role
in social structure.

It must be clear, therefore, that in this "better case of
bifurcation" change characterises the system in its parts and
as a whole although it differs in its character according to
the scale of observation, and also according to the situation
of observation (place and moment in a chronology). Most important, it should also be clear that no part of this system had,
even momentarily, an independent autonomous existence, or that
activation of one or another aspect implies the dormance of all
others until another is selected: we cannot say that now "this
is" and then "that is". All levels and scales, at a given
moment in the chronology of development, exist simultaneously.
Different members of the population are simultaneously operating different criteria, in thought and in practice. Thus, the
metaphor of the computer as memory-bank usefully characterises
a structure of differential possibilities, simultaneously subject to numerous sets of selections, actions and consequences
(take, for example, a telephone-in-computer). The laboratory
conditioned sequence, like the graphically isolable elements of
a severalty of cyclic movements (characteristic of the histories of prices and demography, on which see below), recognises
important moments which tend to be falsely conceptualised in
terms of logical or consequential sequences and oppositions.
The implication is that our "better case" is better not because
it confirms bifurcation, but because it transcends the conditions of its formulation, demonstrating the falsification which
it initially represents.

Now South Asian societies and economies have traditionally
been subject to a systematic body of thinking in which equilibrium, consensus, harmony-oriented, norm-driven practices
(reciprocities), similiar to those observed in "simple" (so-
called "tribal") societies, have a predominant place, a point
which I shall consider in detail below. The characteristic
understanding of Indian social history and social organisation
correlates closely with the ends and ethos of the social
sciences, criticised above, and lacks the kind of well-developed social history, with its method geared to observing
change, characteristic of European history.[63] <u>Jajmani</u>, for
example, is the name given to a nexus of exchanges of goods,
labour and duties of a corporate kind that occur within each
village or group of villages: the whole forms an enclosed set

of rule-dictated, ascriptive reciprocal interactions. The
hierarchies of caste in India, and of lordship in Europe, have
also been treated as actual social universes characterised by
consensus and reciprocal exchange (for which later academic
"apologists" have been legion). But a fully contextualised discussion makes these enclosed options unconvincing.

In this case, then, to speak of equilibrium and non-equilibrium conditions is to lose sight of the complexities and
heterogenous character of the field of motion (its simulta-
neity), and to fail to note that one or another motion, or
scale of motion, only reflects certain abstracted dispositions
in a multi-dispositional reality. In a classic essay on price
movements in Early Modern Europe, Braudel and Spooner describe
the different scales of cyclic and fluctuational movement observed in price movements over the long term (seasonal,
Kitchin's cycle, Juglar's cycle, Labrousses' intercycle,
Baehrel's quadruple cycle, Kondratieff's cycle, to take the
most prominent) and they ask the fundamental question of
whether there is a connection between them, a field of integration, and if so, what form does it take.[64] Their judgement
about the possibility of a final answer (whether, and if so
what) was pessimistic; but we cannot be satisfied with such an
agnostic conclusion since its basis is admittedly a fundamental
ignorance concerning the basic empirical character of these
movements, and ignorance must not be used as an excuse for a
negative theoretical disposition (as it frequently is, although
not in this case by Braudel and Spooner). Our understanding and
description of the part depends on the nature of an answer to
the question: if movements are ultimately autonomous, we can
conceive of closed systems and equilibria - the time of reversible cycles being ostensibly out of time; if, by contrast,
they form a constructional, interdependent "synthesis", equili-
brium-states would be impossible (except as a manner of
speaking, which, however, is best discarded because of the consequences associated with their use and discussed in this
text): "reversibility" would simply be the abstract appearance
of an aspect of what in fact are irreversible structural time-
space continuums.[65] Here we can see quite clearly how the jux-
taposition between equilibrium and non-equilibrium characterises a region of study with notable breaks in knowledge: equilibrium - the closing of a system - forms in the gaps between
the known incidents (between the cycles and their scales),
between, say, the "timeless body" and the "timeful metabolism";
thus the post facto necessity of conceiving transitions of
state from equilibrium to non-equilibrium.

When we look to the historiography of Early Modern Europe,
concerning which Braudel and Spooner were writing, we find a
series of conventional reifications affecting small-scale fluctuations of larger cycles (as in neo-Malthusian historical
demography). Le Roy Ladurie conceives feudal and Early Modern

society as comprising one <u>homeostatic</u> system in equilibrium, a
socio-technological ceiling closing off the possibility of
long-run development – thus a series of long cycles, rises and
falls.[66] Contemporary India and China are the classic cases of
societies treated as systems subjected to long-run cyclical al-
ternations of growth and decline, Raychaudhuri, for example,
describing the late pre-colonial Indian economy as a case of an
immemorial system maintained in a state of "high equilibrium".[67] In every such case – and history is a useful example
in this respect – the field of explanation excludes crucial
sectors of reality. Neo-Malthusianism neglects the problem of
taxation and rents – the general field of appropriation and
exploitation affecting the populations concerned.[68] Stone,
among others, has pointed out the series of similarly neglected
features that enable Le Roy Ladurie to postulate his "homeostatic system". Habib and Raychaudhuri segment India and Europe
into <u>a priori</u> autonomous organisational realities (from category to reality), thereby treating change in Indian society
independently of the growth of the world market, and of European involvement in India.[69] Such notional equilibrium states
depend on failures of synthesis.

It is worth noting how commonly disputations of this type
arise; that between Pédelaborde and Dzerdzeevski over the link
between climatic scales, the debate over causes of ecological
transition, or that concerning essentialism in biology, raise
identical issues.[70]

In summary, the implication seems to be that "bifurcation"
involves an essential mystification, even when used in a merely
figurative sense. It entails a confused identification of a
methodological stance with an actual circumstance, and a philosophical point of departure with an empirical starting point.
Bifurcation is the taming of our philosophical shrew, a seed of
corruption which renders her sterile. In short, we may repeat
that our better case is an unnecessary case (we do not need the
concept), at the same time that it demonstrates the false
premises of the concept (we do not want it).

3.3. Other applications of bifurcation to history

Numerous other examples could be found to which "bifurcation" at first sight appears applicable. Given my criticism
of the presentation of the term, above, this is all the more to
be expected: the diverse kinds of empirical event incorporated
within the term; its barely concealed descriptive rather than
systematic character (similar, indeed, to umbrella-type terms,
such as <u>hysteria, totemism</u> and <u>proto-industry</u>, all of which on
critical examination are shown to consist of diverse, unrelated
and often incomparable kinds of observation).[71] Unfortunately,
many apparently reasonable cases derive from poorly formulated
problems, and outdated conceptualisations, such as mentioned in

the case of counterfactualism. The older manner of putting the
question of historical transition between different types of
society is an example of this kind, traditionally formulated in
terms of an a posteriori process linking two long, separated
epochal stages. Decline in trade between the Roman Empire and
India is thought to have caused far-reaching changes of societal structure, reaching down to village organisation itself,
throughout sub-continental India.[72] A so-called Indian feudalism, of an ideal-typical and static kind, is thereby derived
from abrupt changes of conditions induced by a deus ex machina.
If it is relatively easy to point to the absurdity of this
case, this is because today's historian has access to better,
more complex notions of societal change, but its attraction for
a generation of historians, used to static conceptions of pre-
colonial society, was (and to some extent remains) strong.
There is a powerful analogy between the concept of bifurcation
and this case, in which a view of how change may take place
derives, in the first place, from a priori and static models of
society; both are characterised by dualism between states of
stability and states of change. In the historical case there is
an observable confusion and reification of the typological with
the real (definition of a type, say feudalism, and its transfer
to the prevailing characteristics of a given society over a
period of centuries), but this is only a symptom of the essential problem: the permissive conceptual environment in which
such reification does not seem absurd. It is this close interdependence between a confined, empirical status for irreversibility (both reversibility and irreversibility being allowed to
have separate existences, either in time or space), and a
priori, static conceptions of order, which I shall now examine.
What I wish to emphasise is the underlying primary status assumed of this static order, and the secondary status accorded
to change (thus the corruption of an original philosophy of
irreversibility).

4. SPACE AND ORDER

    The critical problems described above, and which seem to
me to inhibit fulfilment of our current philosophical aspirations (irreversibility, redefinition of the character, content and demarcation of the disciplines), are clearly exemplified in the human sciences. We have an attitude to relevance,
to choice of a site for study, to space and order, which flow
directly from a structure of assumption which contradicts our
more conscious hopes and beliefs. Methodological convenience
cannot be separated from what we consider methodologically
reasonable.[73]
    I pointed out that the treatment of primitive societies,
by anthropologists.... as if they were essentially static sur-

vivals of the past (we can shuffle them into different stages
of an evolutionary sequence, but they lack internal dynamism by
means of which a real transition might occur) - as if they were
isolated in space and also from "vertical", state-influenced
orders (such as under colonial rule) - and as if they composed
a harmonious, cooperative nexus of moral exchanges and duties
(the internal property of a static system, or a system in which
movement is delegated to the diachronic, which comes to the
same thing) ... that all of these characteristics are interdependent with certain methodological dogmas and theoretical assumptions.[74] Since much argument today attempts to controvert
such problematic conceptions and approaches, this schematic
summary cannot be dismissed as oversimplification. I pointed
out that both content and type of data tend to coincide with
the limits of relevance determined by these same stances and
assumptions.[75] A tautological nexus limits the possibility of
truly new departures within the different disciplines, and
this nexus operates at every level, whether that of the coherence of the discipline itself, general theories about the
nature of the subject matter of the discipline, or specific
theories of some aspect or other of the subject matter.[76]

It is in such a framework that the notorious problem of
village and tribal studies are situated. They are not comparable scales of habitation or social order, but they are comparable in other ways. Both are closely related to the basic
ideas characterising the disciplines, as corpuses of knowledge
and a kind of work (cultural anthropology, sociology and economics), and also to the mass of ideas, hypotheses and organisations of specific factual knowledge composing the latter. Given
a disarticulative view of social organisation, and thus attri-
bution of a certain sociological autonomy to the village, and
given the tendency to see peasants and villages as archaic sur-
vivals (a further dimension of disarticulation), it is quite
reasonable to choose the village as a sufficient unit for the
holistic type of study characteristic of anthropology. The
"tribe" is more difficult to pin down as a unit of social organisation of a particular type (not least, because its real
existence is both debated and heavily qualified), but being a
segmentary concept, the tribe itself, or part of the tribe, has
also appeared a suitable unit of study. Tribes and villages, in
both simple and complex societies, are treated alike in this
respect, and despite the vast differences separating pastoral
tribes in the Sudan, say, from villages in the Punjab, they are
seen to possess a similarly autonomous, self-defining organisation, with larger-order contexts (say, states and governments) seen as essentially exogenous.[77] Even in a civili-
sational order of the complexity of India, it remains common to
view the countryside as apart, and in process of penetration by
the larger order or advanced sector. Note the ambiguity, in a
disarticulative perspective, of what is really "apart": the

countryside of villages, a group of villages or the village?
... an insoluble problem characteristic of segmentary thought
systems.[78] Dumont thus describes the very large number of
types of economic activity and exchanges that do not fit into
his jajmani (moral economy) model as deriving from the penetrating, latter-day influences of the modern world outside,
interfering with a formal system of hierarchic organisation and
relationship of quite a different order; symptomatically and
typically, this is without any historical study of what empiri-
cally might have constituted exchanges in the 18th and 19th
centuries, what processes might have been occurring over time,
and without the attempt to phrase questions about the need for
such research and knowledge. The fundamental, underlying assumption is that the countryside is not part of a broader
structure of ongoing relationships that affect the disposition
of what we observe within. What comes from outside is secondary
(until, of course, the occasion of the dissolution of traditional life ascribed to modern influences and deemed responsible
for the contradictions between the empirical matter and one's
model); the primary essentials of village life are ordinarily
autonomous of any outside influence. We thus have autonomous
individuals (albeit corporate) born into the world, between
which relationships of a partial kind are subsequently established.[79]
     Without going into detail about the many theoretical paths
which lead off from this dualistic stance, I wish to concentrate on one of a particularly attractive and influential type,
which has some relevance for this paper: articulation theory.
Dumont's formulation implies that a study of change would need
to seek out those influences from outside which had given rise
to "new" types of activity and relationship (those types not
conforming with the kind characterising the traditional substantivist models). Thus Mendras formulates a theory of traditional villages in 20th century France, being destabilised from
outside by an industrialising society. Between the village and
the state are intermediaries, lawyers, doctors, shopkeepers,
lords, and the like, who articulate and transmit relationships
between the two levels.[80]
     I must now emphasise the substantivist character of this
disposition: the a priori existence of the individual (content
of the reified category); the secondary formulation of the
means of relationship between individuals (between reified
categories). Sociologists and anthropologists find articula-
tionism (a personification of change in the form of activity
and agency residing in a few individuals, who themselves occupy
neither one category nor the other) attractive because of the
substantive, disarticulative assumptions underlying their
thought. Pearson's attempt at sociological analysis of medieval
Gujarat takes the form of a heterogenous medley of different
types and scales of social organisation and institution -

sects, professions, castes, villages, regions, social classes,
national groups – which together form an autonomous bundle of
essentially unrelated collectivities within an explicitly disarticulative sociology.[81] Thus, each group has a formal representative who acts as the intermediary through which relationships between groups need to be conducted (the village headman,
the leader of a sect, etc.). However, we lack studies, of the
"Europeanist" type, of the actual character of relationships
linking the different parts of a societal complex, or capable
of demonstrating that such linkages are absent[82] (although, as
I shall point out below, such study tends to contradict
Pearson's formulations): the hypothesis lacks an empirical base
and indeed is arbitrary in the choices of the possible dispositions of relationships – who and what mediates them, say –
potentially existing between the parts. It may be further
pointed out that the deus ex machina, conjured up to explain a
transition to "feudalism" in India, is just such a kind of a
posteriori articulating mechanism, here temporal, while in the
cases of Mendras and Pearson it is structural (synchronic);
these are logically identical and not surprisingly possess the
same explanatory function.

Among marxist economists and sociologists (chosen here
because of their ostensible stake in dynamic interpretations),
there are identical positions over the concept of mode of production, which for non-historians and sociologists may be described as the attempt to define the meaningful unit of produc-
tion-linked activities capable of replacing itself. Given a
structuralist, pattern-forming approach to the universe (to
take up Prigogine's terms), the mode of production will be defined on the scale of huge assemblages of vertical and horizontal relationships, contexts which, ideally and ultimately,
should lack any conventionally assumed frontiers. But, if you
have a disarticulative view, the mode of production may be defined according to logically limitless choices on a great variety of levels: village, family, tribe, peasantry as a given
generic type, plantation, type of farm, even type of tenancy.
Norman Long conjures up shareholding and plantation modes of
production, and Hans Medick and Marshal Sahlins family modes of
production.[83] Long is very useful because he explains that he
does not consider the Marxist theory of mode of production sufficient, and he summons the sociologist's theory of articulation and intermediaries to fill the gaps between his modes
(firstly the categories, secondly the relationships).[84] This
is not so much a problem with the Marxist theory of the mode,
as such, as with the disarticulative use of it. Thus, mode of
production, in such a perspective (the disarticulative), turns
out to be so positivistic a conception (any formal category,
thus any "term", is potentially a mode) that its application
cannot be limited by either scale or type of body. It exists
independently of, and prior to, any attempt to define struc-

tural relationships. In short, we have a concept as descriptive
in kind (masquerading as an explanatory tool) as bifurcation
threatens to be, and, like bifurcation, it stems from a highly
traditional and a priori view of space and order, despite its
apparently avant garde appearance.

5. PROCEDURAL BIFURCATION. A SATISFACTORY CASE AND A CONCLUSION

A sceptical view (whatever its foundation) is invaluable
in providing means for formulating alternative theories, and
for providing the data necessary to argue them. Only a sceptical view can light up the areas of ignorance which "articula-
tionism" functions to paste over.

With a certain sense of irony, I would describe this reconstruction as a consciously organised expression of a process
of subjective, or procedural bifurcation (entailing the critical discipline of context); it sets off from one state of
knowledge, characterised by stasis and limited scales of order,
and follows a path that enables the constituents to be reinte-
grated into new, larger and patterned structures of a dynamic,
interdependent and unpredictable kind. This a priori interdependence - thus a focus upon relationships and objects as of
equal logical significance - is analogous to the "communications" between molecules (or ants, via deposits of chemical
substance) postulated by the bifurcationists when a change of
order and state occurs. The altered interpretational sophistication resulting from a magnification of scale (from the disarticulation of relatively simple parts to societal contexts of
great organisational complexity), is similar to Deneuborde's
contrast between the macro-structure of the ant's nest and the
micro-structure of an ant's brain, or between the macro-structure of form and the micro-structure of the molecule. It is
also irreversible, because the new research on which such a
reinterpretational stance is based undermines the existential
possibility of the earlier model (logically speaking), and even
demonstrates its pathological character. At the same time it
implies a transition, however gradual and conditional upon deed
and will, of one epistemology into another.

In short, I am suggesting that an invalid objective bifurcation derives from reification of a valid procedural (or subjective) bifurcation, the result of which is that an earlier
epistemology continues to dominate and undermine uneasy attempts to explore beyond it.

I have been able to demonstrate the practical relevance of
these statements in my own research. The kind of substantivist
disarticulation characterising a Dumont (the sociologist's village, the model of jajmani) or a Pearson (pre-colonial India),
is interdependent upon substantial ignorance of areas crucial
either for demonstrating it or attacking it. For example,

Pearson argues that nobles were not interested in managing
their landed possessions, and thus posits the disarticulation
between peasant and the larger order.[85] However, the evidence
of close managerial concern with land is abundant in documentation deriving from a hierarchy of accounting systems; the
latter treats the ordered, exploitative relationships of a
highly differentiated local society. Again, the study of money
provides data of a specifically integrative kind (despite the
existence of disarticulative theories concerning circulation).
The accounts of state and private land-owners demonstrates the
extent to which peasants, craftsmen, soldiers, labourers and
servants were integrated into a money economy which far trans-
cended the locality, and even depended for its supplies of
money (thus for its relative stability and structural characteristics) upon a growing international organisation of trade
and division of labour (Peruvian silver, Japanese copper,
Maldives cauri, Persian badam, the merchants of different na-
tionality and the trade routes which supplied them).[86]

At the same time that such a contextualising approach
permits a different and broader conception of social organisation to be elaborated (and a sceptical dismantling of the
arbitrary limits erected around what is considered relevant),
it also provides the necessary means for reconstructing dynamic
processes within particular local contexts. Macro-structural
organisation and micro-study are closely interdependent in this
sense, a point also made by the bifurcationists.

What I finally wish to suggest, therefore, in controver-
tion to Feyerabend, is that it is indeed possible to chose
between methods and theories.[87] Those approaches which are
impatient of all given boundaries (locating them as a result of
study and not as a condition of it) and which are full-hear-
tedly sceptical, and which thus make a fetish of studying context for context's sake, as well as event, individual and relationship, each for their own sakes, generate by their own
impetus the collection of new kinds of data and new hypotheses,
by means of which disarticulative methodologies and theories
can be tested. All those types of method and theory which derive from substantivist kinds of assumption, generally serve to
confirm given frameworks of thought (because they channel their
approach and search for information within and in terms of the
boundaries of the given categories). Undeniably, theory, method
and data are interdependent, and those who argue that a chosen
method is simply a tool of convenience beg the fundamental
questions. The discipline of context is subversive of objective
bifurcationism; it springs from the very philosophy of irreversibility to which Prigogine's most significant writing attaches.

Acknowledgements

I thank Irfan Habib, J.C. Heesterman, Otto van der Muizen-
berg, Tim Putman and Andre Wink for critical comments on
earlier drafts of this essay and Antoon van den Braembussche
for discussion of some of the ideas raised in it. Two of these
earlier drafts were written in 1974 and 1980, the first for a
seminar at London University ("In search of theory...") and the
second for the 1980 KOTA Conference in the Netherlands. Only in
this version have I accented the argument towards bifurcation,
although I hope that it will be read as a part of the more
general debate of which bifurcation is a part. A number of
issues treated in the present paper have been discussed at
greater length in Perlin, 1984.

NOTES

1. For the earlier case see Lovejoy, 1936: Ch.X (and see 317),
   and for that addressed in this paper, Prigogine, n.d. On
   the former, Cassirer, 1950: Ch. VII-IX, is also useful,
   especially 160ff on the major epistemological difficulties
   involved. The introduction of irreversibility is not the
   same as that of temporality, as is clear from the discussion in Kantorowicz, 1957: 218-232, about the progress from
   an organological conception of society to that of univer-
   sitas. The latter is neo-platonic and deals with entities
   which include time but are unchanging, resting on a separation of the universal from the phenomenal. I deal with
   modern versions of this "static temporality" below.

2. Abrams, 1971 & 1980; but Giddens, 1979: 8, has recently
   reargued the case for a fusion of sociology and history:
   "... History and sociology become methodologically indistinguishable". I had encountered this work after I had completed my paper but have attempted to integrate his position into the footnotes where possible. On a more practical
   level, a similar fate seems to have befallen General
   Degrees and Area Studies in England, popular during the
   '60's but consisting largely of aggregates of disciplines
   instead of new syntheses.

3. Prigogine, n.d. The qualification that functionalists were
   aware of change (Giddens, 1979: 17) may often be true rhe-
   torically but cuts little ice in so far as their methodologies were concerned; this is demonstrated by the actual
   kind of knowledge they built up, as I remark below. This
   was in all essentials static, in its form and in its
   characterisations of its objects of concern, its methodologies reflecting fundamental attitudes to the non-Western
   world.

4. Radcliffe Brown, 1952: Chs. IX & X. Biologism has a genea-
    logy that goes back into the 19th century, but what espe-
    cially marks 20th century sociology and anthropology is
    their rejection of the historicist speculation typical of
    the later 19th century, and thus of the influence of
    theories of evolution carried over from the natural
    sciences. Mendras clearly represents this dilemma with his
    reorientation towards the study of a directly observable
    here-and-now, thus the separation of the synchronic as the
    essential field of inductive, verifiable study (see
    Mendras' two essays in Dion, et al, 1974: 11-58, and my
    review of this book, 1975). Thus the type of biological
    thought used in anthropology was essentially "physiolo-
    gical", or of an essentially static, bounded and function-
    alist kind. In fact, such thinking goes back into the early
    middle ages, when organological models of society (models in
    which the ordering and functioning of the parts of society
    is identified with the assumed ordering and functioning of
    the parts of the human body) were current (besides reference
    to Kantorowicz in note 1, see the frontispiece of Hobbes'
    Leviathan). See also Cassirer, 1950: 99, for the phenomen-
    ologist physicists whose methodological aspirations bear a
    remarkable similarity to those of the anthropologists dis-
    cussed in the text (although of an earlier generation).

5. Lakatos, 1970: 94.

6. Giddens, 1979: Introduction, gives a good recent summary of
    the points raised in this paragraph.

7. Prigogine, n.d.: 2.

8. Prigogine, n.d. 2, quoting Einstein: "For us convinced phy-
    sicists the distinction between past, present and future is
    an illusion, although a persistent one".

9. Lovejoy's (1936) excellent account of this confrontation is
    sufficient to identify this common element; see also Koyre,
    1957.

10. Giddens, 1979, Ch. 1, is especially useful on this separ-
    ation in fields as diverse as anthropology, sociology,
    linguistics and literary criticism. He remarks that "Accor-
    ding to Wittgenstein ... the traditional concerns of Western
    metaphysics have been bound up with the pursuit of illusory
    essences, the search to encompass the 'plenitude of the
    sign'" (34). The distinction is clear in de Saussure's dif-
    ferentiation between langue and parole (de Saussure, 1959:
    14: "In separating language from speaking we are at the same
    time separating: (1) what is social from what is individual;

(2) what is essential from what is accessory and more or
less accidental".); although there is dispute concerning how
representative is this version of de Saussure's thought,
what is significant is its general acceptance and promotion
by structuralist anthropologists, linguists and literary
critics, who replicate this separation. Giddens writes of
Radcliffe Brown and Malinowski (anthropologists) and de
Saussure, that they "come to accentuate the importance of
the 'system', social and linguistic, as contrasted with the
elements which compose it" (9). See also Harris, 1969: 2,
who puts the argument somewhat differently, Lovejoy, 1936:
e.g. 228-291 & 331-332, and a fascinating passage in
Cassirer, 1950: 173 (paraphrasing the views of Christian
Wolff), on the antecedents.

11. Here I am taking history at its own worst image among
    social scientists, qualifying this extreme position below,
    while at the same time emphasising its general isolation
    and peculiar ethos, which protected its special brand of
    empiricism from the series of wrong directions taken by the
    social sciences. It may also be emphasised that if even
    "worst-case" history turns out to have the value I claim
    for the subject in general, my argument is strenghtened.

12. Some of these will be considered below. The "translation"
    of much economic history writing from a "branch of history
    to a branch of economics", is now recognised by many as
    having had just such a retarding effect, a particularly
    serious case since involving a systematic subjection of
    history to the problems and methodologies of one of the
    ahistorical disciplines. The flirtation with anthropology
    is more typical. For a critique see E.P. Thompson, 1972.

13. Giddens. 1979: 3, is again good on the separation of the
    synchronic and diachronic: "The repression of time in
    social history ... is an inevitable outcome of the mainten-
    ance of the distinction between synchrony and diachrony, or
    statics and dynamics, which appear throughout the litera-
    ture of structuralism and functionalism alike".

14. Whilst not completely satisfied with this explanation, it
    has the provisional value of indicating the nature of a
    problem requiring solution, and of correlating well with my
    general hypothesis about context and change. The point is
    further strengthened when history is compared with physics;
    thus according to Cassirer, 1950: 81ff, "Between theoreti-
    cal physics and epistemology there is not only a constant
    reciprocative and uninterrupted collaboration but there
    seems to be a sort of spiritual community" (81); "From the
    middle of the nineteenth century onward the demand for re-

flective criticism in the natural sciences was urged with
ever mounting emphasis" (82). See also Heisenberg, 1962:
18-19 & 34-35, and as paraphrased in Cassirer, 1950: 83.
The contrast with history is obvious.

15. These points concerning bifurcation were made by participants to the conference, but see Prigogine, n.d. & 1971.

16. Lovejoy, 1936: 320 (quoting Oken's Lehrbuch der Naturphilo-
sophie, 1810).

17. Lovejoy, 1936: 320. Gidden's formulation in terms of
"structuration" is strikingly similar to Prigogine's, but
also idential to that of Sartre on which see below
(Giddens, 1979: 54).

18. See for example Gidden's treatment of Derida (1979: 28-33).

19. Given the distinctions made here, it is worth pointing out
that the conference, so it seemed, was marked by a number
of fault lines accross which communications failed. One was
that between quantifiers and non-quantifiers, but much more
serious was that between "philosophizers" and "model
builders". This fault line seems to cut through Prigogine's
own presentation, separating the Jekyll of his philosophy
from the Hyde of his method and hypothesis.

20. See note 3 above. Reciprocity is closely related to equilibria and decontextualised systems. Gouldner, 1960, is typical of its sociological extension to complex societies in
the West; Dumont, 1970, to India; Mauss, 1954, to a specific anthropological problematic; while Sahlins, 1968, un-
wittingly combines it with the complete pathology of properties characterising static conceptual treatments of
"societies". For criticism, see Bourdieu, 1977: 3-9, and
below.

21. This is also to state that the "common sense" of the discipline dictates what methods are reasonable and no method is
reasonable that contradicts the basic dogmas.

22. Examples include Godelier and Meillassoux, the first in his
essay on "tribe" (1977: Ch. III), where, after an excellent
critique of the status of the concept, assumes without
argument its continued value as an anthropological concept;
the second in his functionalist essay of 1973, reducing
time to a series of reciprocal demographic exchanges. The
problem is most obvious in the case of Sahlins (1968 and 1972)
much admired by Godelier (1977: 78ff) who, despite radical
terminology and criticism simply reconstructs the traditio-

nal cultural-substantivist arguments (see also note 21), or
Goody (1976: 21) who, treating western societies as class
societies and thus critically, nevertheless accepts the
common assumption that in India "caste" excludes the possibility of "class".

23. Godelier, 1978: 764, 766 (symptomatically formulated in
less ambiguously culturalist terms by Sahlins, 1976: 20).
What would alter Godelier's formulation from one that is
traditional to one with the revolutionary implications he
apparently favours, is a theory of the acting subject and
structuration, thus of history.

24. Gilbert, 1982, and Trexler, 1981.

25. This basic, apparently contradictory confrontation between
methodologies and stocks of knowledge is clearest when anthropology turns to third world history; thus Dirk's (1979)
study of a pre-colonial state is almost wholly drawn from
19th century colonial experience and completely isolated
from its contexts (neighbouring polities, the colonial
state surrounding it and dictating to it): thus, the traditional Indian village writ large. See also Stern, in Fox,
1977, and Fox, 1971, in which decontextualisation and a
lack of concern to reconstruct actual histories, allows
formulation of a basically static, cyclic theory of Rajput
state formation (thus, as in the case of Dumont, the
"British Government" is required to introduce change).
Geertz's study of a pre-colonial Indonesian state is also
drawn from 19th century experience.

26. Gilbert's remarks are better placed in context with a new
critical literature on the basic validity of a number of
anthropological concepts. A remarkable example is the
debate on the concept of "tribe", to which both anthropologists and historians have contributed (thus see Fried,
1967: 154-174, for an outright dismissal of the concept;
Godelier, 1977: Ch. III, for an ultimately more ambiguous
critical treatment; Southall, 1976, for a particular and
important case (Nuer and Dinka); Colson's remarkable
studies of 1951 and 1953 for the colonial "invention" of
tribal organisation in Africa and the United States;
Ranger's (1977 & 1979) studies of colonial invention in
East Africa (and see the compilation of references used by
him). Ranger, 1979, forms part of the more generally critical literature on the "invention of tradition". Historians
have largely rewritten the terms on which West Africa must
be treated; thus the burgeoning discussion on state-form-
ation and the impact of international trade on hinterland
policy and economy from the 16th century on. A final

example consists of studies of deculturation where "simple" groups and bands living in marginal conditions have been shown to have evolved from much more complex origins under pressure from powerful neighbours and conquerors (see Steward, 1955, and Levi-Strauss, 1963: Ch. VI).

27. For example, Johnson, 1970, effectively dismantles Polanyi's substantivist case for Dahomey (e.g. 1968) by examining a much more extensive empirical base for the spread of money-use.

28. At the time of writing, access to these views is limited to the contributions to the conference, and these form the basis of the following presentation, together with experiments described in Prigogine, n.d.

29. See also Zyman, 1982, for a critical review of Prigogine's book; he also finds the notion of irreversibility fundamental and that of bifurcation unconvincing.

30. Prigogine, n.d.

31. Bourdieu, 1969, for an important criticism, and Levi-Strauss, 1969: 481-482, for the same point in terms of a specific case. The "special solutions" of substantivism may be compared with the problem of essentialism in biology; both act as hindrances to the acceptance of dynamic interpretations. "Essentialism", like substantivism among social scientists, may be thought of as "the automatic unconsidered philosophy of every physical scientist"; see Mayr, 1982.

32. Substantivism denies that exchanges are really economic; they are instead thought to derive from cultural and normative dispositions of a reciprocal or functional, essentially moral, character (see Sahlins, 1968, Ch. V).

33. Prigogine, n.d.: 4.

34. See my essay on "money-use", forthcoming, n. 3, for references.

35. However, these strategic areas of apparently "necessary" ignorance are often complemented by a resistance against accepting that archival records exist which might touch upon them; the phrasing of new questions and the recognition of new kinds of documentation seem to be interdependent.

36. Feyerabend, 1975: Ch. X; see also Cassirer, 1950: 170 on Darwin.

37. See Koyre's two essays "Galileo and Plato", and Galileo and the scientific revolution of the seventeenth century" (in Koyre, 1968), for stimulating discussion of these points and also Cassirer, 1950: 81.

38. These were experimental situations referred to by Prigogine and his colleagues in their contributions to the conference. See Prigogine, n.d.

39. Hobsbawm, 1981: 9.

40. Hobsbawm, 1981.

41. Robinson, 1932; Trotsky, 1932.

42. Sartre, 1963: 90-92 notes 2 & 3, for very clear statements of the theory. Compare "One must understand that neither man nor their activities are in time, but that time, as a concrete quality of history, is made by men on the basis of their original temporalisation", with Oken's statements and with Prigogine's chronotopography, each of which is essentially a fundamental attack on the "'time' of Cartesian rationalism", time as the "universal container" or external measure. Sartre's theory is essentially drawn from the empirical methods and arguments characterising the latter parts of Marx's Capital (1965), in which the problem of the role of the acting subject and structuration, the connection between the levels and scales of action and their structural contradictions, are worked out in a relatively complete manner (not quite so surprising when we consider it as a materialisation of the kinds of formulation expressed by Oken).

43. Sartre's attempts to work out his ideas in terms of historical examples are hardly convincing and thus, much less successful than Marx's; for exemplification of the method it would be better to turn to a practising historian or sociologist such as Bourdieu.

44. Bourdieu, 1977, is the major text, but see also his essay of 1976 for a more accessible working out of the concept of habitus. "To substitute strategy for the rule is to reintroduce time, with its rhythm, its orientation, its irreversibility"; "To abolish the interval is also to abolish strategy" (1977: 6 & 9).

45. Thus see the essays by Berkner, Le Roy Ladurie, and Sabean, in Goody, Thirsk and Thompson, 1976; Berkner's work is especially well-conceived.

46. Malthusianism has had all the omnipotence of a law of
    nature with its querulously obstinate assumption of what
    is common sense. As for the criticism, one can see the im-
    plications in the earlier work of the "inheritance" histor-
    ians, although they seem to retreat from drawing them; see
    for example, Levine, 1977; Medick, 1976, and Plakans, 1975
    among historians and Mamdani, 1972; White, 1976; and Cain,
    1977, among anthropologists.

47. This is the common criticism levelled against the propon-
    ents of the non-Malthusian "demand for labour theory", in
    which relational pressures (such as tax and rent demands,
    occupational possibilities and the price for labour) dic-
    tate familial strategies towards family size.

48. Thus, the family reconstruction method used by Levine,
    1977, and which as originally formulated was an attempt to
    penetrate behind poor census material and use particular
    cases for building up a generalisation of general popul-
    ation trends; it has been remarkably successful. Note that
    this is a clear example of the virtues of comparison and
    interdisciplinarianism, argued in this essay: each new link
    in the chain — each destruction of a boundary previously
    separating one area of explanation from another — improves
    the coherence, power and quality of the explanation. Metho-
    dologically, the comparison of new trends in third world
    demography with European historical demography, improves
    the quality and comprehensiveness of the hypothesis: thus
    Medick, 1976, explains the "demand for labour theory" as a
    special case characterising rural industrialisation in the
    18th century, but similar phenomena observed in neo-feudal
    Kurland, also in the 18th century, Java and present day
    Bangladesh, show the problem to be of more general inci-
    dence.

49. It may also be in terms of the closure of a district, or
    group of villages, or of a polity (assumed as sufficient
    for explanation); see note 25 for examples.

50. Unfortunately, Sartre, 1976: 125, distinguishes societies
    "which are nevertheless correctly classified by ethno-
    graphers as societies with no history, societies based on
    repetition", which "have begun to interiorise our history,
    because they have been subjected to colonialism as a his-
    torical event". This slip hardly affects Sartre's central
    argument, since it is not the basis of his methodology as
    is the case with Prigogine (thus bifurcation). It is
    another example of the ambiguous attitudes towards stocks
    of knowledge incompatable with consciously formulated phi-
    losophies. These are examples of the frequently observed

manner in which discredited ideas are allowed new homes in
"peripheral" regions of the world, or peripheral sectors of
knowledge (what I elsewhere call "residualism").

51. For example, Deyon and Mendels, 1981: 13-14 ("En effet, les
conséquences de la proto-industrialisation furent première-
ment de rompre le système auto-régulateur de la démographie
ancienne"); Braun, 1978: e.g. 331-332; Levine, 1976: 178.
Thus another example of the "residualism" treated in n. 50.

52. Compare Turner's criticism of Piaget's refusal to isolate
any discrete level of reality apart from the genetic, mistakenly explained by a puzzled Turner (1973: 361-362, cri-
ticising especially Piaget, 1968), as due to a kind of intellectual stubbornness. According to Turner, "The variable
and potentially independent relationship between genetic
and self-maintaining or homeostatic processes would seem to
imply the possibility of a purely synchronic "construc-
tivism", based upon the cyclical or self-regulating process
of systems considered as maintaining themselves in equilibrium over virtual or relatively short periods of time,
which could dispense with 'diachronic' (historical, developmental, and genetic) considerations. It would also
leave room for a diachronic (historical) but non-genetic
structuralism, that would concern itself with temporal modifications in structural systems resulting from conflict,
maladaptation to changing circumstances, contacts with
other systems, or other forms of relative 'disequilibrium',
without binding itself to any overall scheme of genetic
development or evolution". Synchrony, on the one hand,
equilibrium, on the other, are precisely the reserve areas
where the old epistemology is stored for its residual and
convenient purposes. Similarly in linguistics, "Diachrony
is held by Jacobson to produce inbalances that lead to re-
adjustments on the level of synchrony, hence connecting the
two" (Giddens, 1979: 28).

53. An example also used by Hobsbawm, 1981, our argument being
similar at this point. I thank C. Lis for this reference.

54. Thompson, 1972.

55. Dzerdzeevskii, 1963: especially 285 (and see the useful
discussion in Le Roy Ladurie, 1971: 303-304, which does
not, however, convey the power of Dzerdzeevksii's own
statement).

56. On space and time, see note 65.

57. The following section is based on my own research.

58. Note that although I describe contemporaries as (in part)
    substantivists, I am not applying a substantivist explan-
    ation; moreover, I do not regard substantivism as a proper-
    ty of particular kinds of society and not others but, in
    combination with other very different tendencies of thought
    and action, as a universal, thus also of the modern indus-
    trial world. I discuss this whole issue at length in my
    interim paper of 1984; see also note 62. Praxis, in my
    text, means the unity (but not similarity, which implies
    dualism) of thought and practice comprising a body of
    knowledge, the continuity between knowledge and institution,
    and between knowledge and practical action. On the problem-
    atic historical relationships between reference systems and
    idealogies, again see my above cited paper.

59. Ullman, 1966, gives an excellent account of "descending"
    types of ideology.

60. This is clearly so in the case of heritable property; prop-
    erty is the origin, logically speaking, of the civic per-
    sona, while lack of property often implies non-recognis-
    ability, social invisibility (as with Locke's labouring
    class). But this is also the case in a hierarchic ideology
    where to belong to a category headed by another is to be
    recognisable. See Pocock's (1979) valuable essay for dis-
    cussion of some of these points.

61. These statements derive from my own research. The ident-
    ification of ideological pluralism (including egalitarian-
    ism) would seem to necessitate a basic change in the manner
    in which we interpret pre-colonial orders.

62. Writing of the 1950s and 1960s, Giddens, 1979: 28, writes
    that "Structures were usually in that period treated as
    given codes, examined within closed and discrete systems".
    It is the same kind of thinking as presented in modern
    examples of substantivism and criticised in Bourdieu, 1969;
    it is to be explained but not used as an explanation (which
    would be to accept the given ideology as sufficient).

63. See Said, 1978, on "orientalism", being as relevant for
    India as for the Middle East, with which Said is mainly
    concerned.

64. Braudel and Spooner, 1967: especially 430-442 (noting that
    the existence of some of these cycles is disputed). Pomian.
    1979: 639, remarks that "taken together, the ... tendencies
    ... mentioned seem to give human history a linear, cumula-
    tive and irreversible character. One might be tempted to
    draw the conclusion that the good old idea of human

progress has not fared so badly..."; note that it is this
use of irreversibility, in the form of Social Darwinism and
other distortions of evolutionary theory, which have discredited much 19th century social-science thinking. However, Pomian rightly goes on to remark that irreversibility
no longer implies a given value-laden teleology.

65. This "analogy" between Dumont's decontextualised jajmani
    and the disarticulation of different time-scales implies
    that the problem of space and time in this respect is one
    problem. Thus, we could say that the relationship between
    the village and its larger context is equivalent to that
    between small-scale fluctuations and larger movements, con-
    textualisation introducing the irreversible (for apparently
    epistemological reasons, differences between space and time
    seem to disappear at the limits of astronomical observ-
    ation, a point implied by Dzerdzeevskii (n. 55)). In con-
    trast, Prigogine, 1971: 12: "Clearly we perceive a whole
    hierarchy of structures separated by discontinuities".

66. Le Roy Ladurie has formulated this concept in numerous
    works but see his short article of 1972 for a clear formu-
    lation. Once again, his opponents seem to formulate their
    views according to similar principles; thus according to
    Hobsbawm, 1965: 15, "unless certain conditions are present
    ... the scope of capitalist expansion will be limited by
    the general prevalence of the feudal structure of society
    ...". An underlying conjunction of views among disputants
    is nicely put by Cassirer: "Often both thesis and anti-
    thesis stood on the same ground". (Cassirer, 1950: 86-87,
    of certain disputes in theoretical physics); Giddens, 1979:
    1, remarks that "there are no easy dividing lines between
    Marxism and 'bourgeois social theory'" (more acurately,
    despite contrasting aspirations, epistemological obstacles
    tend to weaken these dividing lines, as we have seen with
    the dissident anthropologists. Thus, Le Roy Ladurie, 1978:
    56, with some justice replies to an attack on his "homeo-
    static" theory with "The fact that Guy Bois considers
    himself a Marxist only makes my own argument the stronger",
    which, however, is not to criticise Bois for attacking the
    theory. Real advance has to find a means to break through
    this conspiracy of tacit agreements, which means facing the
    problem at an appropriate level.

67. In characterising the fate of the Mughal Empire and its
    failure to industrialise, Habib, 1969: 50, cites Mao Tse
    Tung's (Selected Works, III, 1954: 75-76) view of Chinese
    history as subject to recurrent cycles. According to Ray-
    chaudhuri, 1982: 307, late pre-colonial India had an eco-
    nomy which constituted "an immemoral system functioning at

the level of a high equilibrium". Fox, 1977, also profers a
theory of recurrent cycles, and there are many others. See
Perlin, 1983, for an alternative approach.

68. Neo-Malthusians such as Le Roy Ladurie and Goubert provide
a good deal of valuable data on these aspects but fail to
integrate it into their models of causation.

69. Stone, 1979, with respect to Le Roy Ladurie; on comparti-
mentalisation of Indian and European histories, Perlin,
1983.

70. See references in note 55, and Pédelaborde, 1957/1958.

71. For "hysteria", Foucault, 1965: 138; for "totemism", Levi-
Strauss, 1962: 1-14, for "proto-industrialisation", Perlin,
1983; jajmani seems to be a similar catch-all notion.

72. Thus Sharma, 1974.

73. The following section is based on my contribution to the
KOTA conference on "Region", Rotterdam, 1980, where I
argued against the view that an increase in the scale of
the unit studied by social scientists, would automatically
solve the problems of contextualisation.

74. In essence, evolutionists were often no less subject to
static approaches than their counterparts. A neo-evolu-
tionist like Sahlins, 1968, provides no internal means
whereby his "tribes" might alter or evolve; his stone age
and neolithic societies. like those of the 19th century
anthropologist, involve an arbitrary shuffling of "peoples"
existing today into an invented chronological ordering.
Thus the importance of critiques such as Southall, 1976;
Levi-Strauss, 1963, Ch. VI, and other references in n. 26.

75. Thus Stein, 1980, whose important synthesis fails to consider the kind of information that might have contradicted
his particular substantivist interpretation of medieval
South India (for example, use of money and taxation systems).

76. There is more than a mere analogy between the problem of
compartimentalised "disciplines" and "regions"; it is pecu-
liarly exemplified by the manner in which Althusser takes
the disciplines as given and conceives them as "regions", a
significant terminological ambiguity characterising the
French language. And see Cassirer, 1955, I: 77, for similar
"reification" of the disciplines. The present-day interdisciplinary temper makes both appear archaic in this respect, yet not much has really changed.

77. Thus, the study of simple, acephalous societies, such as the Nuer of the Sudan in the mid-1930s, has provided stimulus for a theory of segmentary political organisation in the case of the complex states of medieval South India (Evans Pritchard, 1940; Stein, 1980).

78. See Kantorowicz, 1957: 210, 304-305, for corporate fictive persona in medieval Europe; I am researching such persona in late pre-colonial Maharashtra; in both cases the idea seems easily transferable from one socio-institutional scale to another, and from one institutional body to another.

79. Williams, 1973, for these dualisms on the popular and literary level; they seem to be the "natural way" in which the countryside is customarily perceived (see also n. 31).

80. Mendras, 1974, portrays a mosaic of villages, disarticulated in both time and space.

81. Pearson, 1976: especially Ch. VI.

82. I use the awkward term "societal complex" in place of "society" in order to bypass a notion that is too often confused with bounded, complete entities.

83. Long, 1977: Ch. IV; Medick, 1976: 304 and section II for assumptions; Sahlins, 1972 (who uses the term "domestic mode of production").

84. Long, 1977: Introduction, Chs. IV & V.

85. Pearson, 1976: Ch. VI.

86. On the relationship between international supplies of monetary media and developments in hinterland economies, see Perlin, forthcoming (money-use).

87. Feyrabend, 1975, argues for an unqualified relativism, "anything goes", in order to generate maximum flexibility.

REFERENCES

Abrams, Philip: 1971, Sociology and History, Past and Present,
LII, pp. 118–125.

Abrams, Philip: 1980, History, sociology, historical sociology,
Past and Present, LXXXVII, pp. 3–16.

Berkner, Lutz, K.: 1976, Inheritance, land tenure and peasant
family structure: a German regional comparison, in: Jack
Goody, Joan Thirsk and E.P. Thompson (eds.), Family and inheritance. Rural society in Western Europe 1200–1800, Cambridge University Press, Cambridge, pp. 71–95.

Bourdieu, Pierre: 1968, Structuralism and the theory of sociological knowledge, Social Research, XXXV, pp. 682–706.

Bourdieu, Pierre: 1976, Marriage strategies as strategies of
social reproduction, in: Robert Foster and Orest Ranum
(eds.), Family and society. Selections from the Annales
E.S.C., Johns Hopkins University Press, Baltimore and London
pp. 117–144.

Bourdieu, Pierre: 1977, Outline of a theory of practice, Cambridge University Press, Cambridge.

Braudel, F.P. and Spooner, F.: 1967, Prices in Europe from 1450
to 1750, in: E.E. Rich and C.H. Wilson (eds.), The Cambridge
Economic History of Europe, Volume IV, The economy of expanding Europe in the 16th and 17th centuries, Cambridge
University Press, Cambridge.

Braun, Rudolf: 1978, Early industrialization and demographic
change in the canton of Zurich, in: Charles Tilly (ed.),
Historical studies of changing fertility, Princeton University Press, Princeton.

Cain, Mead T.: 1977, The economic activities of children in a
village in Bangladesh, Population and Development Review,
III, pp. 201–227.

Cassirer, Ernst: 1950, The problem of knowledge. Philosophy,
science and history since Hegel, Yale University Press, New
Haven and London.

Cassirer, Ernst: 1955, The philosophy of symbolic forms, 3 volumes, Volume I, Language, Volume II. Mythical thought, Volume III, The phenomenology of knowledge, Yale University
Press, New Haven and London.

Colson, Elisabeth: 1951, The Plateau Tonga of Northern Rho-
desia, in: Elizabeth Colson and Max Gluckman (eds.) Seven
tribes of British Central Africa, Oxford University Press,
London.

Colson, Elisabeth: 1953, The Makah Indians, University of Min-
nesota Press, Minneapolis.

Deyon, Pierre and Mendels, Franklin: 1981, Programme de la Section A2 du Huitième Congrès International d'Histoire Econo-
mique: la proto-industrialisaton: théorie et réalité, Revue
du Nord, LXIII, pp. 11-19.

Dirks, Nicholas, B.: 1979, The structure and meaning of political relations in a South Indian little kingdom, Contributions to Indian Sociology, New Series, XIII, pp. 169-206.

Dumont, Louis: 1970, Homo Hierarchicus. The caste system and
its implications, Weidenfeld and Nicholson, London.

Dzerdzeevskii, B.L.: 1963, Fluctuation of general circulation
of the atmosphere and climate in the twentieth century,
Changes of Climate. Proceedings of the Rome Symposium,
Unesco, Paris, 285-296.

Evans-Pritchard, E.E.: 1940, The Nuer. A description of the
modes of livelihood and political institutions of a Nilotic
people, Clarendon Press, Oxford.

Feyrabend, Paul: 1975, Against method. Outline of an anarch-
istic theory of knowledge, NLB, London.

Foucault, Michel: 1965, Madness and civilization. A history of
insanity in the age of reason, Vintage Books, New York.

Fox, Richard, G.: 1971, Kin, clan, raja and rule: state hinterland relations in pre-industrial India, University of California Press, Berkeley.

Fox, Richard G. (ed.): 1977, Realm and religion in traditional
India, Vikas Publishing House, New Delhi.

Fried, Morton, H.: 1967, The evolution of political society. An
essay in political anthropology, Random House, New York.

Giddens, Anthony: 1979, Central Problems in social theory. Action, structure and contradiction in social analysis, The
Macmillan Press, London.

Gilbert, Felix: 1982, The Medici Megalopolis, New York Review of Books, XXVIII, nos. 21-22, pp. 62-66.

Godelier, Maurice: 1977, The concept of the "tribe": a crisis involving merely a concept or the empirical foundations of anthropology itself?, in: Maurice Godelier, Perspectives in Marxist anthropology, Cambridge University Press, Cambridge, pp. 70-96.

Godelier, Maurice: 1978, Infrastructures, societies, and history, Current Anthropology, XIX, 763-768.

Goody, Jack: 1976, Production and reproduction. A comparative study of the domestic domain, Cambridge University Press, Cambridge.

Gouldner, Alvin W.: 1960, The norm of reciprocity: a preliminary statement, American Sociological Review, XXV, pp. 161-178.

Habib, Irfan: 1969, Potentialities of capitalistic development in the economy of Mughal India, Journal of Economic History, XXIX, pp. 32-78.

Harris, Marvin: 1969, The rise of anthropological theory. A history of theories of culture, Routledge and Kegan Paul, London.

Heisenberg, Werner: 1962, La nature dans la physique contemporaine, Editions Gallimard, Paris.

Hobbes, Thomas: undated reprint of 1651 edition, Leviathan or the matter, forme and power of a commonwealth ecclesiastical and civil, edited with an introduction by Michael Oakeshott, Basil Blackwell, Oxford.

Hobsbawm, E.J.: 1965, The crisis of the seventeenth century, in: Trevor Aston (ed.), Crisis in Europe 1560-1660, Essays from Past and Present, Routledge and Kegan Paul, London, pp. 5-58.

Hobsbawm, Eric: 1981, Looking Forward: history and the future, New Left Review, 125, pp. 3-19.

Johnson, Marion: 1970, The cowrie currencies of West Africa, Journal of African History, pp. 17-49 & 331-353.

Kantorowicz, Ernst H.: 1957, The king's two bodies. A study in mediaeval political theology, Princeton University Press, Princeton.

Koyre, Alexander: 1957, From the closed world to the infinite universe, Johns Hopkins University Press, Baltimore and London.

Koyre, Alexander: 1968, Metaphysics and measurement. Essays in scientific revolution, London, pp. 16-43.

Le Roy Ladurie, Emmanuel: 1972, Population and subsistence in sixteenth century rural France, Peasant Studies Newsletter, pp. 60-65.

Le Roy Ladurie, Emmanuel: 1976, Family structures and inheritance customs in sixteenth-century France, in: Jack Goody, Joan Thirsk and E.P. Thompson (eds.), Family and inheritance. Rural society in Western Europe 1200-1800, Cambridge University Press, Cambridge, pp. 37-70.

Le Roy Ladurie, Emmanuel: 1978, Symposium: Agrarian class structure and economic development in pre-industrial Europe. A reply to Professor Brenner, Past and Present, LXXIX, pp. 55-59.

Lakatos, Imre: 1970, Falsification and the methodology of scientific research programmes, in: Imre Lakatos and Alan Musgrave (eds.) Criticism and the growth of knowledge, Cambridge University Press, London, pp. 91-196.

Levine, David: 1976, The demographic implications of rural industrialization: a family reconstitution study of Shepshed, Leicestershire, 1600-1851, Social History, I, pp. 177-196.

Levine, David: 1977, Family formation in an age of nascent capitalism, Academic Press, London and New York.

Levi-Strauss, Claude: 1963, The concept of archaism in anthropology, in: Claude Levi-Strauss, Structural Anthropology, Penguin Books, Harmondsworth, pp. 101-119.

Levi-Strauss, Claude: 1964, Totemism, Merlin, London.

Levi-Strauss, Claude: 1969, The elementary structures of kinship, Eyre & Spottiswoode, London.

Long, Norman: 1977, An introduction to the sociology of rural development, Tavistock Publication, London.

Lovejoy, Arthur O.: 1936, The great chain of being. A study of the history of an idea, Harvard University Press, Cambridge, Massachusetts and London.

Mamdani, Mahmood: 1972, The myth of population control. Family,
    caste, and class in an Indian village, Monthly Review Press
    New York and London.

Marx, Karl: 1965 reprint, Capital, A critique of political eco-
    nomy, Volume I, Capitalist Production, translated from the
    third German edition by Samuel Moore and Edward Aveling,
    edited by Frederick Engels, Lawrence & Wishart, London.

Mauss, Marcel: 1954, The gift. Forms and functions of exchange
    in archaic societies, Routledge and Kegan Paul, London.

Mayr, Ernst: 1982, The growth of biological thought: diversity,
    evolution and inheritance, Harvard University Press,
    Cambridge, Massachusetts and London.

Medick, Hans: 1976, The proto-industrial family economy: the
    structural function of household and family during the tran-
    sition from peasant society to industrial capitalism, Social
    History, I, pp. 291-315.

Meillassoux, Claude: 1973, The social organization of the pea-
    santry, Journal of Peasant Studies, I, pp. 81-90.

Mendras, Henri: 1974, Un schéma d'analyse de la paysannerie
    francaise, and: Schémas d'analyse villageoise, in: Michel
    Dion, Nicole Eizner, Marcel Jollivet, Jacques Maho and Henri
    Mendras, Les collectivités rurales francaises, Tôme II,
    Société paysannes ou lutte de classes au village. Problèmes
    methodologique et théoriques de l'étude locale en sociologie
    rurale, Armand Colin, Paris, pp. 11-58.

Pearson, M.N.: 1976, Merchants and rulers in Gujerat: the res-
    ponse to the Portuguese in the sixteenth century, University
    of California Press, Berkeley.

Pédelaborde, Pierre: 1957-1958, Le climat du bassin parisien.
    Essai d'une methode rationelle de climatologie physique, 2
    Volumes, Paris.

Perlin, Frank: 1975, (Review of Michel Dion et al, 1974), Jour-
    nal of Peasant Studies, II, pp. 245-249.

Perlin, Frank: 1983, Proto-industrialization and precolonial
    South Asia, Past and Present, xcviii, pp. 30-95.

Perlin, Frank: 1984, (forthcoming), Money-use in late pre-colo-
    nial India and the international trade in currency media,
    revised version of a paper presented to the Conference on
    the Monetary System of Mughal India, Duke University.

Perlin, Frank: 1984, Exchange-economy and culture in late pre-
colonial India: an alternative model, to appear in the proceedings of the South Asia Regional Seminar 1983/84: Merchants, Capital and Commerce, University of Pennsylvania.

Piaget, Jean: 1968, Structuralism, Routledge and Kegan Paul,
London.

Plakans, Andrejs: 1973, Peasant families east and west: a com-
ment on Lutz K. Berkner's 'Rural family organization in
Europe': a problem in comparative perspective, Peasant
Studies Newsletter, II, pp. 11-16.

Pocock, J.G.A.: 1979, The mobility of property and the rise of
eighteenth-century sociology, in: Anthony Parel and Thomas
Flanagan (eds.), Theories of property. Aristotle to the
present, Wilfrid Laurier University Press, Waterloo, pp.
141-166.

Polanyi, Karl: 1968, Archaic economic institutions: cowrie
money, in: George Dalton (ed.), Primitive, archaic and
modern economies. Essays of Karl Polanyi, Beacon Press, Boston, pp. 280-305.

Pomian, Krzysztof: 1979, The secular evolution of the concept
of cycles, Review, II, pp. 563-646.

Prigogine, Ilya: 1971, Unity of physical laws and levels of
description, in: Marjorie Grene (ed.), Interpretations of
life and mind. Essays around the problem of reduction, Routledge and Kegan Paul, London.

Prigogine, Ilya: 1980, From being to becoming, W.H. Freeman.

Prigogine, I.: nd, Probing into time, The A. Katzir-Katchalsky
Lecture.

Prigogine, I., Allen, P. and Deneubourg, V.: January 1982, Introduction to bifurcation theory, lectures given to the Col-
loquium on Bifurcation Theory, Erasmus University Rotterdam.

Radcliffe-Brown, A.R.: 1952, Structure and function in primitive society. Essays and addresses, Cohen and West, London.

Ranger, Terence: forthcoming, The invention of tradition in
Colonial Africa, paper presented to the Past and Present
Conference 1977: The invention of tradition, London Univer-
sity.

Ranger, Terence: 1982, Race and tribe in Southern Africa. Euro-
    pean Ideas and African acceptance, in: Robert J. Ross (ed.),
    Colonialism and racism, Martinus Nijhoff, The Hague, pp.
    121-142.

Raychaudhuri, Tapan: 1982, Non-agricultural production. I.
    Mughal India, in: Tapan Raychaudhuri and Irfan Habib (eds.),
    The Cambridge Economic History in India, Volume I: c1200-
    c1750, Cambridge University Press, Cambridge, pp. 261-307.

Robinson, Geroid, Tanquary: 1932, Rural Russia under the Old
    Regime. A history of the landlord-peasant world and a pro-
    logue to the peasant revolution of 1917, The Macmillan Com-
    pany, New York.

Sabean, David: 1976, Aspects of kinship behaviour and property
    in rural Western Europe before 1800, in: Jack Goody, Joan
    Thirsk and E.P. Thompson (eds.), Family and inheritance.
    Rural society in Western Europe 1200-1800, Cambridge Univer-
    sity Press, Cambridge, pp. 96-111.

Sahlins, Marshall D.: 1968, Tribesmen, Prentice Hall, Englewood
    Cliffs.

Sahlins, Marshall: 1972, Stone age economics, Tavistock Publi-
    cations, London.

Sahlins, Marshall: 1976, Culture and practical reason, Univer-
    sity of Chicago Press, Chicago and London.

Said, Edward: 1978, Orientalism, Routledge and Kegan Paul,
    London & Henley.

Sartre, Jean-Paul: 1968, Search for a method, Vintage Books,
    New York.

Sartre, Jean-Paul: 1976, Critique of dialectical reason I.
    Theory of practical ensembles, edited by Jonathan Ree, NLB,
    London.

Saussure, Ferdinand de: 1959, Course in general linguistics,
    The Philisophical Library, New York.

Sharma, R.S.: 1974, Problems of transition from Ancient to
    Mediaeval in Indian history, Indian Historical Review, I,
    pp. 1-9.

Southall, Aidan: 1976, Nuer and Dinka are people: ecology, eth-
    nicity and logical possibility, Man, New series, II, pp.
    463-491.

Stein, Burton: 1980, Peasant state and society in Medieval South India, Oxford University Press, New Delhi.

Stern, Henri: 1977,Power in traditional India: territory, caste and kinship in Rajasthan, in: Richard G. Fox (ed.), Realm and region in traditional India, Vikas Publishing House, New Delhi.

Steward, Julian H.: 1955, Theory of culture change. The methodology of multi-linear evolution, University of Illinois Press, Urbana.

Stone, Lawrence: 1979, (review of Le Roy Ladurie, Carnival in Romans), New York Review of Books, XXVI, No. 17.

Thompson, E.P.: 1972, Anthropology and the discipline of historical context, Midland History, pp. 41-55.

Trexler, Richard C.: 1981, Public life in Renaissance Florence, Academic Press, New York and London.

Trotsky, Leon: 1932, The Russian Revolution. The overthrow of tzarism and the triumph of the Soviets, selected and edited by F.W. Dupree, from: The History of the Russian Revolution, Doubleday Anchor Books, New York.

Tung, Mao Tse:, 1954, Selected Works, Volume III, London.

Turner, J.: 1973, Piaget's Sructuralism, American Anthropologist, 1973, LXXV, pp. 351-373.

Ullmann, Walter: 1966 (2nd edition), Principles of government and politics in the Middle Ages, Methuen and Co., London.

White, Benjamin N.F.: 1976, Production and reproduction in a Javanese village, doctoral thesis, Columbia University.

Williams, Raymond: 1973, The country and the city, Chatto & Windus, London.

Zyman, John: 1982, Irreversibility, London Review of Books, IV, No. 5, (review of Ilya Prigogine, From being to becoming).

C. Synthesis

SYMMETRY, BIFURCATIONS AND PATTERN FORMATION (d' apres Sattinger, Michel, Thom and many others)

Michiel Hazewinkel

Econometric Inst., Erasmus University, Rotterdam and
Centrum voor Wiskunde en Informatica (= Centre for
Mathematics & Computer Science), Amsterdam

1. INTRODUCTION

Nature often seems to like (approximately) symmetric solutions to problems. Mathematically, or more generally, scienti-
fically, it thus becomes our task to understand why, e.g. by
showing that more or less regular patterns are usually the most
stable, or economical, or optimising with respect to a suitable
criterion.
Particular aspects of this theme concern the question of
whether symmetric problems necessarily have symmetric solutions
and how the symmetry of a problem/solution can change as a
parameter varies. This last problem in turn is part of bifurcation theory which examines the question of how the set of
solutions of a problem can change in nature as certain parameters vary. Here it turns out to be remarkably fruitful [9, 29,
33] to take the symmetry of a situation into account. Also the
presence of a symmetry group has remarkably strong consequences
for extrema of functions as we shall see.
There are two aspects of symmetry which I consider very
interesting and which perhaps so far have not had all the attention they deserve. One concerns approximate symmetry. Here I
have in mind for instance a symmetric problem which would have
a symmetric solution if the boundary conditions were equally
symmetric. Now suppose the boundary conditions are disturbed,
when will there be an approximatedly symmetric solution, e.g.
when will there be a boundary layer in which the symmetry will
be restored and a fully symmetric solution in the middle or
when will the solution of the new problem be like a crystal
with defects. There seems to be no general theorem concerning
such matters. Other approximate symmetry problems (e.g. how to

M. Hazewinkel et al. (eds.), Bifurcation Analysis, 201–232.
© *1985 by D. Reidel Publishing Company.*

recognise them) arise when dealing with a basically symmetric
pattern with a (small) random [3] or systematic disturbance
superimposed. Some preliminary foundational remarks on approximate symmetry are contained in [17, 28]. Much more remains to
be done and different kinds of approximate symmetry certainly
exist. Still harder to understand approximatedly symmetric entities are the Penrose universes described in section 2 below.
The second aspect concerns the matter that as certain parameters change an object may both lose and gain symmetry, sometimes simultaneously. The systematics of symmetry loss, that is
spontaneous breaking of symmetry, have had considerable attention over the years, see e.g. [17, 20, 21, 22, 24]. The systematics of gaining symmetry far less. The matter is discussed
below in section 3 in terms of the automorphisms of three dimensional algebras.

Apart from this example there is little new in this paper
and it should basically be seen as a low key introduction to
important work of others, notably the persons mentioned in the
title. I have added a few more references than are strictly
necessary for the purposes of this paper itself, e.g. a few
references to books which treat of bifurcation theory and give
applications [32, 25, 18, 26, 11, 1] and a few (in my view)
closely related matters [2, 10, 41, 23].

2. PATTERN FORMATION AND SYMMETRY AND APPROXIMATE PATTERNS AND SYMMETRIES

One of the best known examples of pattern formation and
one of the most studied (the two terms are not synonymous) is
the Benard convection. Consider a fluid layer heated from below
as in Figure 1. At a small temperature difference (small temperature gradient) heat is transported by conduction and the
solution

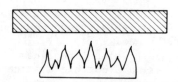

Figure 1. Fluid layer heated from below. After [12]

is completely symmetric, that is if we consider only the possible symmetrics viewable from the top we have E_2 symmetry
where E_2 is the group of rigid motions of the plane (the fluid
layer is assumed to be infinitely extended). At higher temperature gradients the lighter fluid at the bottom will tend to
rise, cool at the top and return to the bottom. A microscopic
pattern arises which can take the form of rolls, c.f. Figure 2

or the famous Bénard cells (hexagons), c.f. Figure 3.

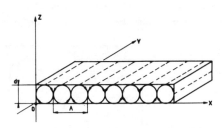

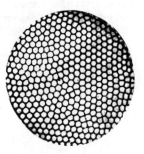

Figure 2. Rolls pattern.   Figure 3. Hexagonal
After [12]   pattern. After [5]

Other less regular patterns develop at still higher temperature
gradients. We will say a little more about Sattinger's bifurca-
tion-in-the-presence-of-symmetry analysis of the Bénard problem
in section 5 below. A highly recommended up to date discussion
of the topic is [9], which includes also a discussion of the
spherical case, a model for convection in the molten layer of
the earth between core and mantle.

Figure 4. East wall of Death Valley, California.
After [35].

Another example of a strikingly regular pattern, caused by erosion in this case, is depicted in Figure 4 and still more examples are the drainage basin patterns of the Figures 5, 6, and 7 below. These drainage patterns seem perhaps less regular

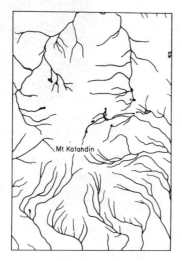

Figure 5. Radial drainage pattern. After [39]

Figure 6. Parallel drainage pattern. After [39]

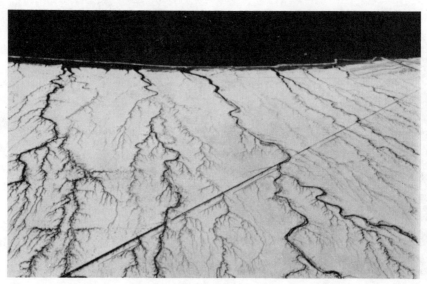

Figure 7. Dendritic channels. After [35].

than the erosion pattern of Figure 4 and the convection patterns of the Figures 2 and 3. Still they are very far from random and as such demand an explanation of their (amount of)
regularity. In all these examples the problem is not so much to
understand that something happens, i.e. that some pattern de-
velops, but to explain the striking regularity of the patterns
and in case there are several patterns to understand the selection mechanisms and the relative stability of these patterns
with respect to each other.

Still other examples of spontaneously arising regular patterns are the socalled cloud streets, c.f. Figure 8 below, and
the socalled Liesegang rings, which form e.g. when a drop of
silver nitrate is placed on a film of gelatine saturated with
potassum dichromate as in Figure 9.

However, both nature and man seem to like approximately
symmetric solutions even better. Or solutions whose obvious
regularity is much harder to describe in mathematical terms
then e.g. hexagonal or street patterns. Spiral patterns for
instance, occur very frequently, c.f. Figures 10 and 11 below

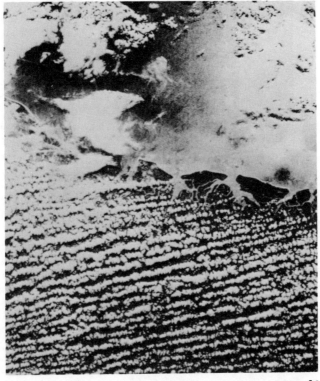

Figure 8. Stratocumulus cloud streets. After [37]

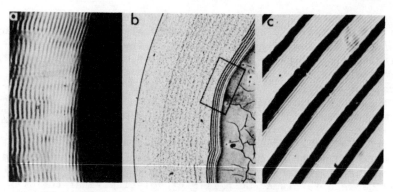

Figure 9. Liesegang rings. After [4]

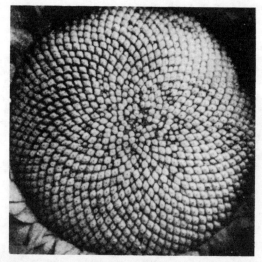

Figure 10. Sunflower head. Courtesy of Empire magazine

Interesting remarks on the mathematics of spirals can be found in for instance [6]. Still harder to describe kinds of symmetry are those exhibited by various fractal patterns such as the twin dragon pattern of Figures 12 and 13 below which not only have certain more or less obvious symmetries (c.f. Figure 13) and the less obvious rotational symmetry of Figure 12, also has the property that it can be covered with reduced size replicas of itself ad infinitum.

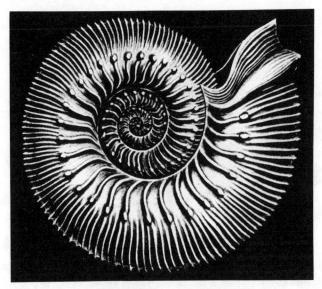

Figure 11. After [36]

Figure 12. Twin dragon fractal, After B.B. Mandelbrot,
The fractal geometry of nature, Freeman, 1982

Consider also the spiral tiling of H. Voderberg depicted
in Figure 14. Though it obviously has many regularity properties and is intuitively very symmetric it is quite difficult to
find symmetries (apart from a 180 degree rotation which also
interchanges colours).

Figure 13. Tiling of twin dragon fractal, After B.B.
Mandelbrot, The fractal geometry of nature, Freeman,
1982

Figure 14. A non-periodic tiling by H. Voderberg. From
[8]

There are patterns whose "regularity" is even more distur-
bing and very hard, possibly impossible to describe in the mathematical framework we usually use in this connection. These
are the Penrose non-periodic tilings, also called the Penrose
universes as described in [8] from which the following is
taken.

The basic tiles are obtained from a diamond with angles of
72° and 108° as drawn in Figure 15. The number t is the golden
ratio $\frac{1}{2}(1 + \sqrt{5})$. The two tiles, called dart and kite, so obtained are marked with a drawn and dashed circle as indicated
and there is an additional tiling rule in that abutting edges
must join circle segments of the same kind so that fitting a dart
and a kite to form a diamond is forbidden. Using these two kinds

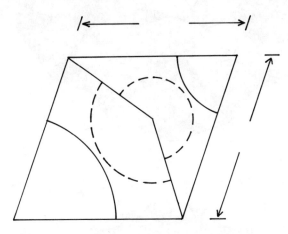

Figure 15. The Penrose dart and kite.

of tiles and respecting the additional tiling rules it is possible to tile the entire plane. Indeed there are innumerable
different ways of doing that. Some of the more striking patterns are depicted below in the Figures 16, 17 and 18. The central 10-sided regular polygon consisting of 15 darts and 25
kites in Figure 18 is called a cartwheel. Note also that the
cartwheel pattern of Figure 18 has little symmetry in the obvious sense (only a reflection through a central vertical
line).

Here are some properties of the Penrose tilings (or universes):
a) all tilings are non-periodic;
b) every point in every tiling is inside a cartwheel;
c) every finite region of diameter ≤ d of any tiling occurs
within distance 2d of any point in any other tiling.

In addition there are "fractal properties" in that from a
given tiling another one with larger kites and darts can be
constructed in a simple systematic way.

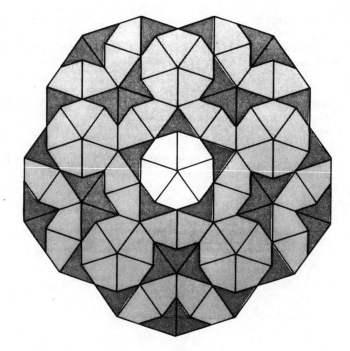

Figure 16

Figure 17

Figure 18

Properties b) and c) above are regularity, indeed symmetry, properties especially property c) when applied to the same
tiling. The "fractal property" is also a symmetry property of
course. Yet these symmetries are of a different kind then what
we usually understand by the word.

2. CHANGES IN SYMMETRY AS A PARAMETER VARIES

This section contains a number of remarks pertaining to
what can happen to the symmetry group of an object as a parameter varies, i.e. during a deformation.

2.1. Example

Consider a rectangle with sides 1 and λ For $\lambda \neq 1$, the symmetry
group is generated by the two reflections across the central
horizontal and vertical axis, so that the symmetry group is the
Klein four group $V_4 = Z/(2) \times Z/(2)$. For $\lambda = 1$ however, there
suddenly appears an additional bit of symmetry, viz a rotation
through 90°. For this value of λ the symmetry group is suddenly
larger.

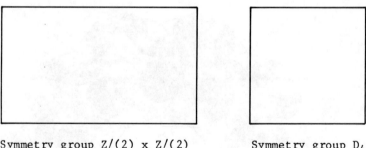

Symmetry group Z/(2) x Z/(2)   Symmetry group D_4
Figure 19

2.2. Example

    Consider a small square inside a larger one with the sides
of the small square parallel to those of the big one as in Figure 20. The parameter which varies is the position of the smaller square

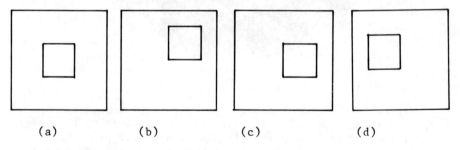

(a)   (b)   (c)   (d)

Figure 20

If the centre of the small square is precisely in the middle of
the big one the symmetry group of the whole figure is D_4 (Figure 20 (a)), if the centre is on a diagonal but not in the centre the only non-trivial symmetry is a reflection across that
diagonal (Figure 20 (b)). The situation is analogus for the
centre on the horizontal or vertical symmetry axis of the big
square (Figure 20 (c)) and finally if the centre is on none of
these four lines there is no non-trivial symmetry. Thus, in the
parameter square of this example we can describe the symmetry
of the various figures as in Figure 21 below. A similar picture
holds for example 2.1. And indeed in certain reasonably general
situations one can show that this is the general pattern as we
shall see below in section 2.3. Though of course it may happen
that at a certain critical value the symmetry group increases
by an infinite amount as when one considers the rigid

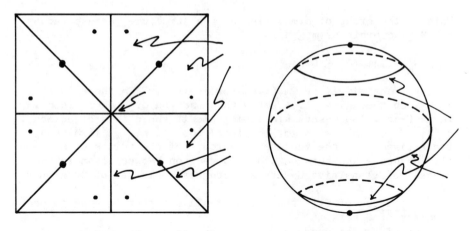

Figure 21   Figure 22

motions of the plane which take a given ellipse with major axis
λ and minor axis 1 into itself. When λ becomes one (circle) the
symmetry group suddenly increases from $Z/(2) \times Z/(2)$ to the
circle group S^1 of all rotations about the centre of the ellipse.

2.3. A general mathematical framework for symmetry breaking [20, 21]

    We can view the examples above as follows. There is a
group of potential symmetries, in this case the rigid motions
of the plane. For certain isolated parameter values a "large"
subgroup of this group defines actual symmetries and as the
parameter moves away from this value the symmetry is broken to
a smaller subgroup.
    A more precise and mathematical setting for this picture
of symmetry breaking is as follows.
    Let G be a group acting on a set M. (I.e. there is given a
map $G \times M \to M$, written $(g,m) = gm$ such that $g_1(g_2 m) = (g_1 g_2)m$
$1m = m$; think of gm as the result of applying the transformation g to m). For instance in the case of example 2.2. we
would have $G = E_2$, the group of rigid motions of the plane, and
the set M is the set of all squares of sizes 6 and 2, the latter inside the other with sides parallel to the co-ordinate
axes of the plane. M can be conveniently described as M
$= \{(P_1, P_2) \quad R^2 \times R^2 : d(P_1, P_2) \leqslant 2\}$ (P_1 is the center of the
first square, P_2 that of the second).

    For each $m \in M$, the isotropy subgroup at m is

$$G_m = \{g \in G : gm = m\} \tag{2.1}$$

This is the group of symmetries of the structure represented by
m   M. The _orbit_ of m $\in$ M is

$$G_m = \{gm : g \in G\} \tag{2.2}$$

G acts transitively on Gm (i.e. for every $x,y \in$ Gm there is a
$g \in G$ such that gx = y). A set with an action of G on it is called a G-set. Two G-sets are isomorphic if there is a bijection
$\phi: M_1 \to M_2$ such that $\phi(gm_1) = g\phi(m_1)$ for all $m_1 \epsilon M_1$. If M is a
transitive G-set the isotropy subgroups of the points of M are
all conjugate and this sets up a bijection between isomorphism
classes of transitive G-sets and conjugacy classes of subgroups
of G.

A _stratum_ of G is the union of all orbits belonging to one
isoclass of G-sets.

Consider for example M = S^2, the sphere, and G = $O_2(R)$ the
group of all rotations around the N-S axis and inversion
through the origin. An orbit is then the union of two parallel
circles at equal northern and southern latitude. There are
three strata viz {N} $\cup$ {S} and the equator, both are strata
consisting of a single orbit, and the third stratum is the
union of all other orbits. A precisely similar picture is offered by the description of example 2.2. by means of Figure 21.
By restricting immediately to the subgroup D_4 of E_2 which
leaves the larger square invariant we have G = D_4 and M is a
square of side 4.

Orbits are e.g. sets of 8 points as indicated by dots in
Figure 21, or sets of four points as indicated by crosses, or
sets of four points as indicated by small circles and finally
the centre is an orbit. There are four strata given precisely
by these four types of orbits.

Returning to the general situation. If G is a compact Lie
group acting smoothly on a smooth manifold M everything is
beautiful: the isotropy subgroups are closed Lie subgroups, orbits and strata are submanifolds and there exists a G-invariant
Riemannian metric on M. A consequence of all this is:

2.4. Theorem

Let G be a compact Lie group acting smoothly on a smooth
manifold M. Then for every m $\in$ M there is a neighbourhood U of
m such that G_m is larger than $G_{m'}$ (up to conjugacy) for all
m' $\in$ M.

Thus symmetry can suddenly decrease but not suddenly increase which is precisely as in examples 2.1. and 2.2. and also
as in the example of Figure 22.

Note that in this setting hidden symmetry may occur. To
see this consider the following modification of the example of
Figure 22. Consider again S^2 and let the group G now consist of
rotations around the N-S axis and reflexion through the equator

plane. For M take the space of all unordered pairs of parallels. There is an obvious induced action of G on M by viewing G
as a group of transformations on S^2 so that each element of G
takes an element of M to a possibly different element of M. Let
P be the element of M consisting of twice the equator. The isotropy of P is all of G. The rotations are visible. However, the
reflection (when restricted to the submanifold of S^2 represented by P) acts just like the identity. That is as a symmetry of
the figure P in S^2 it is a hidden symmetry, which appears as
soon as P is deformed into a pair of close together equal latitude north and south parallels.

2.5. Example

Consider the group of all transformations $R^2 \rightarrow R^2$ of the
plane into itself of the form $(f,g) : (x,y) \rightarrow (f(x), y + g(x))$
where $f: R \rightarrow R$ is a diffeomorphism and $g: R \rightarrow R$ is any differentiable map. The inverse of (f,g) is the element $(f^{-1}, -gf^{-1})$
and the composition goes as follows: $(f_1,g_1) \circ (f_1,g_1) = (f_2 f_1, g_1 + g_2 f_1)$. The identity element is (id, o). Thus G is a subgroup of the group $Diff(R^1)$ of all diffeomorphisms of R^2 into
itself. Now let M be the space of all unordered pairs of elements of R^2. Consider an element $P = \{x_1, y_1), (x_2, y_2)\} \in M$ and
let us calculate the isotrophy subgroup of P. Suppose $\phi \in G$ is
in G_p. Then we must have

$$\phi(x_1, y_1) = (x_1, y_1) \text{ and } \phi(x_2, y_2) = (x_2, y_2) \tag{i}$$

or

$$\phi(x_1, y_1) = (x_2, y_2) \text{ and } \phi(x_2, y_2) = (x_1, y_1) \tag{ii}$$

or both which can only happen if $(x_1, y_1) = (x_2, y_2)$ It is
quite easy to describe the isotropy subgroup for all P. For the
purposes of this example, however, we need only two cases.

$x_1 = x_2$ and $y_1 \neq y_2$.

$$\text{(a)}$$

Then $G_p = \{(f,g) : f(x_1) = x_1, g(x_1) = 0\}$

$x_1 \neq x_2$ and $y_1 \neq y_2$

Then $G_p = \{(f,g) : f(x_1) = x_1, f(x_2) = x_2, g(x_1) = \quad\text{(b)}$

$\qquad = g(x_2) = 0\} \cup$

$\quad \{(f,g): f(x_1) = x_2, f(x_2) = x_1, g(x_1) = y_2 - y_1,$

$\qquad g(x_2) = y_1 - y_2\}$

In case (b) the set

$N_p = \{(f,g) : f(x_1) = x_1, f(x_2) = x_2, g(x_1) = g(x_2) = 0\}$ is a

normal subgroup of G_p and G_p is in fact the disjoint union N_p

αN_p where $\alpha = (f,g)$ is any element of G such that

$f(x_1) = x_2, f(x_2) = x_1 , g(x_1) = y_2 - y_1 , g(x_2) = y_1 - y_2.$

Now let $P = ((x_1,y_1), (x_2,y_2))$ approach a point Q

$= ((x,y_1), (x,y_2))$ with $y_1 \neq y_2$. Then we see that as the limit
point is reached there is both symmetry gain in that the N_p
part of G_p becomes bigger and sudden symmetry loss in that the
αN_p part of G_p disappears. It is in fact easy to show that
there are no inclusion up to conjugacy relations between G_p and
G_Q.

Thus theorem 2.6. does not hold in more general situations. It seems that this kind of phenomenon cannot happen
when considering the symmetry group (of motions) of figures or
patterns in Eudidean space as these figures change. But I do
not know of a general theorem to this effect.

3. DESIGN SYMMETRY VERSUS GENERIC SYMMETRY [14]

In example 2.7. we saw that in more general cases a statement like that of theorem 2.6 does not hold. A more complicated
but also more suggestive example of the same phenomenon of simultaneous symmetry loss and symmetry gain during a deformation
is obtained by considering the automorphisms of three (or
higher) dimensional algebras over R. That is the subject of
this section.

3.1. Algebra structures

Let $V = R^3$ be the vectorspace of all 3-tuples of real numbers. An associative algebra structure with unit on V is given
by a bilinear map (the multiplication).
    m: $V \times V \to V$, $(x,y) \to xy$

such that $(xy)z = x(yz)$ and such that there exists a $1 \in V$ with
$1x = x1 = x$ for all $x \in V$. By choosing a basis in V suitably we
can assume that $1 = e_1 = (1,0,0)$ and we shall do so. Let e_2, e_3
be the other basis elements. Then because of the bilinearity
the multiplication is specified by 12 constants (the so-called
structure constants)

$$e_2 e_2 = \gamma_{22}^1 e_1 + \gamma_{22}^2 e_2 + \gamma_{22}^3 e_3 \qquad e_2 e_3 = \gamma_{23}^1 e_1 + \gamma_{23}^2 e_2 + \gamma_{23}^3 e_3$$

$$e_3 e_2 = \gamma_{32}^1 e_1 + \gamma_{32}^2 e_2 + \gamma_{32}^3 e_3 \qquad e_3 e_3 = \gamma_{33}^1 e_1 + \gamma_{33}^2 e_2 + \gamma_{33}^3 e_3$$

In order that the algebra be associative these γ_{ij}^k have to satisfy certain relations. E.g. from $e_2(e_2 e_3) = (e_2 e_2)e_3$ one obtains

$$\gamma_{23}^2\gamma_{22}^1 + \gamma_{23}^3\gamma_{23}^1 = \gamma_{22}^2\gamma_{23}^1 + \gamma_{22}^3\gamma_{33}^1$$
$$\gamma_{23}^1 + \gamma_{23}^3\gamma_{23}^2 = \gamma_{22}^3\gamma_{33}^2 \tag{3.1}$$
$$\gamma_{23}^2\gamma_{22}^3 + \gamma_{23}^3\gamma_{23}^3 = \gamma_{22}^1 + \gamma_{22}^1\gamma_{23}^3 + \gamma_{22}^3\gamma_{33}^3$$

The precise form of these conditions will not be important for us.

3.2. Isomorphisms and automorphisms

A map $\phi: A \to B$ from an algebra A to an algebra B is an isomorphism if it is an isomorphism of vectorspaces and if moreover $\phi(1_A) = 1_B$ and $\phi(xy) = \phi(x)\,\phi(y)$ for all $x,y \in A$. Here 1_A and 1_B denote the unit elements of A and B. An isomorphism $\phi: A \to A$ is called an automorphism and Aut(A) denotes the group of automorphisms of A. This group should also be considered as the group of symmetries of the algebra A. Indeed the situation fits the general framework described above. Let G be the group of all vector space isomorphisms $\phi: V \to V$ such that $\phi(e_1) = e_1$ (because we only consider algebra structures on V for which e_1 is the unit element). Let M be the space of all 12-tuples

$(\gamma_{22}^1, \ldots, \gamma_{33}^3)$ such that all associativity relations like

(3.1) hold. For $\phi: V \to V$, $\phi(e_1) = e_1$, let $\bar{e}_2 = \phi(e_2)$, $\bar{e}_3 = \phi(e_3)$

and let $\bar\gamma_{22}^1, \ldots, \bar\gamma_{33}^3$ be defined by

$$\bar{e}_i \cdot \bar{e}_j = \bar\gamma_{ij}^1 e_1 + \bar\gamma_{ij}^2 \bar{e}_2 + \bar\gamma_{ij}^3 \bar{e}_3 \quad , \quad i,j = 2,3$$

The group element ϕ now acts on M by
$$(\gamma_{22}^1, \ldots, \gamma_{33}^3) \to (\bar\gamma_{22}^1, \ldots, \bar\gamma_{33}^3).$$

One sees immediately that if A is an associative algebra with structure constants $\gamma = (\overset{1}{\gamma_{22}}, \ldots, \overset{3}{\gamma_{33}})$ then $Aut(A) = G_\phi$, the

isotropy subgroup of γ M.

Two algebras A and B are isomorphic if and only if they (or more precisely the corresponding 12-tuples of structure constants) are in the same orbit.

3.3. The isomorphism classes of three dimensional algebras

It turns out that up to isomorphism there are six different algebra structures on $V = R^3$. They are

$A_1 \simeq R[X]/X(X-1)(X-2)$

$A_2 \simeq R[X]/X(X^2 + 1)$

$A_3 \simeq R[X]/X^3$

$A_4 \simeq R[X]/X^2(X-1)$

$A_5 \simeq R[X,Y]/(X^2, Y^2, XY)$

A_6 with basis 1, e_2, e_3 and the multiplication rules
$$e_2^2 = 1, \quad e_3^3 = 0, \quad e_2 e_3 = e_3, \quad e_3 e_2 = -e_3$$

Here if $f(X) = X^3 + a_2 X^2 + a_1 X + a_o$ is a polynomial $R[X]/f(X)$

denotes the 3 dimensional algebra with basis 1, X, X^2 and multi-

plication rules $XX = X^2$, $XX^2 = X^2X = -a_2X^2 - a_1X - a_o$. The al-

gebra A_5 is defined similarly.

3.4. The deformation/contraction relations between the six isomorphism classes

By the symbol $A \Longrightarrow B$ we understand that there is a family

of algebra structures $A(t) = (\overset{1}{\gamma_{22}}(t), \ldots, \overset{3}{\gamma_{33}}(t))$ isomorphic for small $t \neq o$ to A and such that A(o) is isomorphic to B. In other words $A \Rightarrow B$ means that the orbit corresponding to the isomorphism class B is in the closure of the orbit corresponding to isomophism class A.

With this notation the pattern of contraction/deformation relations between $A_1, \ldots, A_6$ is as follows

A_1
A_6
A_4
A_3
A_5
A_2

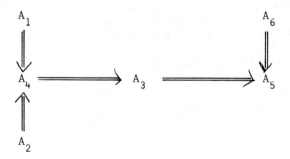

Figure 23

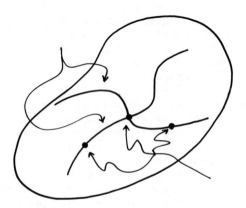

Figure 24

Concentrating on the A_1, A_3, A_4 part of Figure 23 the situation in γ-space is somewhat like depicted in Figure 24. Compare this e.g. with Figure 22 and Figure 21.

The corresponding pattern of automorphism groups is as follows

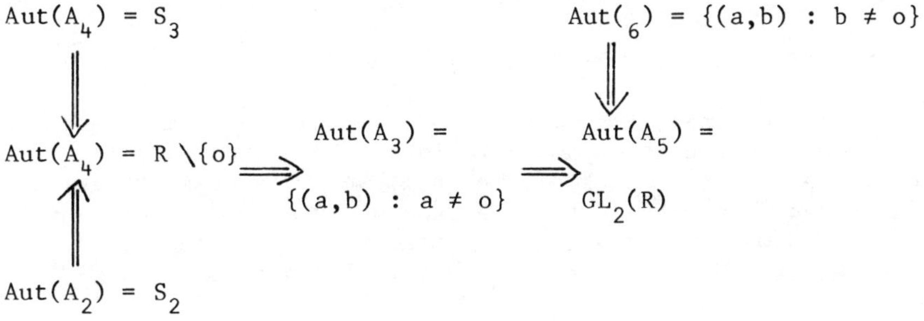

$\mathrm{Aut}(A_4) = S_3$
$\mathrm{Aut}(_6) = \{(a,b) : b \neq o\}$
$\mathrm{Aut}(A_4) = R \setminus \{o\}$
$\mathrm{Aut}(A_3) =$
$\{(a,b) : a \neq o\}$
$\mathrm{Aut}(A_5) =$
$GL_2(R)$
$\mathrm{Aut}(A_2) = S_2$

Figure 25

Here S_n is the permutation group on n letters and $GL_m(R)$ is the group of real invertible m x m matrices. The multiplication rules of $Aut(A_3)$ and $(Aut(A_6)$ are respectively $(a,b)(c,d) =$

$(ac, ad + bc^2), (a,b)(c,d) = (c+ad, bd)$.

We see that as a rule during a contraction ($\twoheadrightarrow$)
(i)   symmetries are both gained and lost
(ii) the dimension of the symmetry group does not become
      less

3.5. Design versus generic symmetry

At least in a large number of cases, there seem to be two sources of symmetry. The first is what I like to call "generic symmetry" it is the symmetry which is possessed by almost all of the structures under consideration. As an example consider algebras of the form $C[X]/f(X)$ where C denotes the complex numbers and $f(X)$ is a polynomial of degree n. Almost all $f(X)$ have n distinct roots and as a consequence almost all of these algebras have S_n as their automorphism group. The second source is what I like to call design symmetry which arises e.g. when parts of the structures under consideration are very carefully arranged in such a way that a large symmetry group arises. For instance if precisely two roots of $f(X)$ are equal and all others are different from each other and from this double root, then the automorphism group picks up a factor $R-\{0\}$ but not all roots are of the same kind anymore and the generic symmetry group drops to S_{n-2} (the permutations of the (n-2) unequal single roots). It seems to me that the way the symmetry groups can change often can be understood systematically in these terms. During contraction ($\Rightarrow$) the generic symmetry group tends to become smaller and the design symmetry group larger or, equivalently, during a deformation ($\Leftarrow$) the generic symmetry group becomes larger and the design symmetry group smaller (i.e. gets broken). This last phenomenon was of course the subject matter of section 2 above.

4. CONSEQUENCES OF THE PRESENCE OF SYMMETRY

Quite generally the presence of symmetry in a (mathematical) problem can have enormous influence and it can greatly facilitate solving a problem. We shall see this when examining bifurcation phenomena in the presence of symmetry in the next section and we have already seen examples in the previous section. Here we describe some more material around this theme.

4.1. Symmetric problems and their solutions

A first question to examine is whether symmetric problems necessarily have symmetric solutions. This is discussed in considerable detail by W.C. Waterhouse in [40] who calls the principle that this be the case the Purkiss principle. Here are some of his examples where the principle holds.
- of all rectangles with a given perimeter the square has the largest area.
- Take four positive numbers whose product is 16. Then their sum is least when all numbers are equal.
- For a given mean $\bar{x} = n^{-1} (x_1 + \ldots + x_n)$ the value of $x_1^2 + \ldots + x_n^2$ is least when all x_i are equal.

It is clear that statements to the effect that under certain circumstances the Purkiss principle holds are precisely the desired sort of mathematical explanations of why nature likes symmetric solutions (Cf. the introduction).

There are also quite simple counter examples to the Purkiss principle. For instance one from Buniakovsky: find the minimum of $f(x,y) = (x^2 + (y-1)^2)((x-1)^2 + y^2)$. This is symmetric in x and y. The two solutions, however, are (1,0) and (0,1). Another example (from [20]) which I like very much is the following. Consider four towns located on the corners of a square. What is the shortest road system that joins these four towns. It is not very difficult to see that there are two solutions which are depicted in Figure 26. The angle between the horizontal and the top left oblique segment in Figure 26 is 30°. This solution is better than the other obviously possible candidates: three edges of the square or the two diagonals.

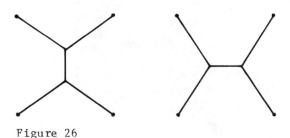

Figure 26

Note that in both counter examples the total set of solutions is symmetric under the full symmetry group of the problem but that the individual solutions (if there is more than one) are only invariant under an isotropy subgroup. That is, there is symmetry breaking in precisely the sense of section 2 above with as the manifold M the space of all solutions.

4.2. Extrema of symmetric functions

Here is a result that shows how strong the influence of
the presence of non-trivial symmetry can be. The setting is
that of section 2.3 above, i.e. a compact Lie group G (for instance a finite one) acting smoothly on a smooth manifold M.
Let F be the set of all functions f on M which are invariant
under G, i.e. such that f(gm) = F(m) for all m ε M, g ε G

Theorem (cf. e.g. [21]). If an orbit is isolated in its
stratum it is critical for all f Є F (i.e. df = 0 at all points
of that orbit) and inversely if an orbit is critical for all
f Є F it is isolated in its stratum.

5. BIFURCATION IN THE PRESENCE OF SYMMETRY

5.1. General remarks and first examples

Bifurcation theory is concerned with how the set of solutions of a problem can change as a parameter varies. For a
first introduction to bifurcation theory I refer the reader to
my chapter "Bifurcation phenomena. A short introductory tu-

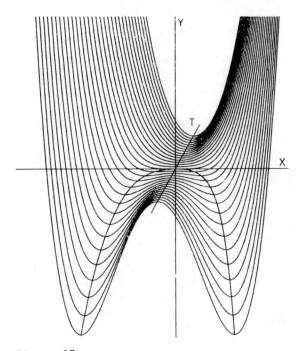

Figure 27

torial with examples" in this volume [15]. Two of the simplest
examples of bifurcations are the pitchfork bifurcation depicted
in Figure 27 where a minimum bifurcates into two minima and a
maximum, and the Hopf bifurcation where a stable equilibrium
point bifurcates into an oscillatory cycle. In the case of the
bifurcating valley there is also obviously a kind of symmetry
breaking involved and in the case of the Hopf bifurcation also
seems to involve symmetry loss when viewed in space-time space,
cf. fig 28 below. The continuous translations symmetry gets
broken into a discrete group of translation symmetries. Quite
generally it seems clear from the setting of sections 2 and 4
above that if a solution of a problem is symmetric with
symmetry group G, and for parameter values $\lambda < \lambda_0$ there is
(locally) a single solution, and if at λ_0 this solution
bifurcates into several, then the new solutions will have as
symmetry group isotropy subgroups of G. This should severely

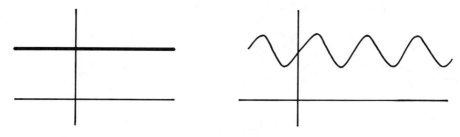

Figure 28

restrict the possible bifurcations. To see how this works, how
the results of section 4 apply, and how representation theory
can be used to advantage we first need a slightly more precise
description of a little bit of bifurcation theory.

5.2. Microsynopsis of some bifurcation theory

Consider a map $G(\lambda,.) : B_1 \rightarrow B_2$ depending on a parameter λ
(where λ may also denote a vector of parameters). Here the B_i
are Banach spaces. At this stage it does not hurt to think of
the B_i as finite dimensional spaces, say $B_1 = B_2 = R^2$ and to
think of $(\lambda,.)$ as a map given by an expression like $(x^3 + y^3 +
\lambda xy^2, x^2 + \lambda xy)$. We are interested in studying $G(\lambda,u) = 0$, $u \in$
B_1, and in how the solution set of this equation changes as λ
varies. In most of the interesting examples the B_i are suitable
spaces of functions and $G(\lambda,.)$ is e.g. a differential operator
depending smoothly on λ. An explicit example occurs below.

Suppose $G(\lambda_0, u_0) = 0$; consider the partial derivative
$G_u(\lambda_0, u_0)$ at (λ_0, u_0). This is a linear map $B_1 \rightarrow B_2$. If this
linear map is invertible then the implicit function theorem
says that there is a differentiable mapping $\lambda \rightarrow u(\lambda)$ for λ near

λ_0 such that $G(\lambda, u(\lambda)) = 0$ and that locally (i.e. for (λ, u)
close to (λ_0, u_0) this is the only solution. Thus for bifurcation phenomena we are interested in points where $G_u(\lambda_0, u_0)$ is
not invertible. In the finite dimensional case of $B_1 = B_2 = R^n$
we are thus interested in cases where the $n \times n$ matrix of
partial derivatives $G_u(\lambda_0, u_0)$ is not of full rank. If there is
a symmetry group involved it is obvious how (generalised) results like those of section 4.2 above could, indeed will be
important.

Assume that $G_u(\lambda_0, u_0)$ is a Fredholm operator of index
zero, so that the kernel is finite dimensional and the range is
closed of co-dimension equal to the dimension of the kernel
(automatically the case if $B_1 = B_2 = R^n$.). In the simplest case
dim $N = 1$, where $N = \text{Ker } G_u(\lambda_0, u_0)$. Let $B_1 \subset B_2$ and let $P : B_2 \to$
N be the projection on N and write $Q = \text{Id} - P$. Then we can rewrite the equation $G(\lambda, u) = 0$ as

$$QG(\lambda, v + \psi) = 0 \quad \text{and} \quad PG(\lambda, \psi) = 0$$

where $v = Pu$, $\psi = Qu$. Fix v, then $QG_u(\lambda_0, u_0)$ is an isomorphism,
so by the same implicit function theorem used above we can find
$\psi(\lambda, v)$ as a function of (v, λ) so that $Q(\lambda, v + \psi(\lambda, v)) = 0$
near (λ_0, v_0). Thus it remains to solve the so-called bifurcation equations

$$F(\lambda, v) \equiv PG(\lambda, v + \psi(\lambda, v)) : R \times N \to N, \quad F(\lambda, v) = 0$$

In the simplest case (dim $N = 1$) it readily follows that $F(\lambda, v)$
is of the form

$$F(\lambda, v) = a(\lambda - \lambda_0) v + \dots$$

so that in the non-degenerate case, the solution set near $(\lambda_0, 0)$ looks like two lines crossing each other vertically, a
pitchfork bifurcation.

5.3. Equivariant bifurcation theory

Now suppose that there is a group of symmetries involved.
I.e. there is a group H acting linearly on B_1 and B_2 and $G(\lambda, u)$
is equivariant which means that $G(\lambda, gu) = g G(\lambda, u)$ for all g
G. Here is a general result [33, theorem 13].

Theorem. Let $G(\lambda, u) : B_1 \to B_2$ be analytic and equivariant
w.r.t a compact group H. Let $G(\lambda_0, u_0) = 0$, $gu_0 = u_0$ for all
g H, and let $G_u(\lambda_0, u_0)$ be Fredholm of index zero with kernel
N_0. Then N_0 is invariant under H (i.e. gN_0 N_0 for all g H)
and the $F(\lambda, v)$ are equivariant.

Now in many interesting bifurcation problems dim $N > 1$.
Then the simple analysis of 5.2 above does not apply. However,

if there is symmetry it may easily happen that N is an irredu-
cible representation of H and if we know which one (i.e. as a
rule, if we know enough of the representation theory of H) this
is just as good as the case dim N = 1.

We also know a priori that the bifurcating solutions will
have symmetry groups which are isotropy subgroups of H acting
on N. A very simple example of a bifurcation situation with
symmetry (rotational symmetry in this case) is the one of a
stiffish rubber bar with opposite forces acting on the two end
points as shown in Figure 30. The corresponding bifurcation
diagram is sketched of Figure 29.

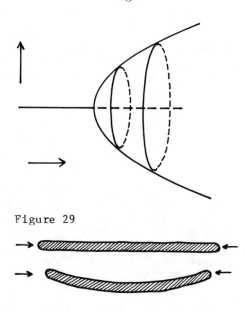

Figure 29

Figure 30

The idea is the following. Just as in case dim N = 1 we
know that $F(\lambda,v)$ must start off with a term λv (if $\lambda_0 = 0$), we
know in the equivariant case that $F(\lambda,v)$ must start off with
very specific terms which are determined by the representation
of H, which is involved. In many cases degree considerations
and the value of dim N rule out all but a few possible (known)
representations which may make a case like dim N = t (which is
usually totally intractable in the general case) quite easy to
do in a symmetric case.

5.4. Example: Bénard convection [33, 34]; cf. also [9]

In case of the Bénard convection such an analysis has actually been carried out by Sattinger, loc. cit. The equations
involved are the Boussinesq equations which are

$$\Delta u_k + \delta_{k3}\,\theta - \frac{\partial p}{\partial X_k} = P_r^{-1}\sum_j u_j \frac{\partial u_k}{\partial X_j} \quad , \; k = 1, 2, 3$$

$$\Delta\theta + Ru_3 = \sum_j u_j \frac{\partial\theta}{\partial X_j}$$

$$\sum_j \frac{\partial u_j}{\partial X_j} = 0$$

Here u_1, u_2, u_3 are the velocity components of a fluid element as functions of the co-ordinates X_1, X_2, X_3. The function θ is the temperature profile and P is the pressure. R denotes the Rayleigh number, P_1 the Prandtl number and Δ denotes the Laplacian

$$\frac{\partial^2}{\partial X_1^2} + \frac{\partial^2}{\partial X_2^2} + \frac{\partial^3}{\partial X_3^2}$$

and δ_{k1} the Kronecker delta. The bifurcation parameter λ is R. Thus in this case B_1 and B_2 are suitable spaces of 5-tuples of functions $(u_1, u_2, u_3, \theta, P)$ of three variables X_1, X_2, X_3, and it is then immediately clear how to write these equations in the form $G(\lambda, u) = 0$.

The group involved is E_2, the group of rigid motions of the plane which consists of all motions

$$g\colon \begin{pmatrix} X_1 \\ X_2 \end{pmatrix} \to \begin{pmatrix} a_1 \\ a_2 \end{pmatrix} + \begin{pmatrix} b_{11} & b_{12} \\ b_{21} & b_{22} \end{pmatrix}\begin{pmatrix} X_1 \\ X_2 \end{pmatrix}$$

where the matrix is orthogonal and hence of determinant ± 1.
The action of E_2 on the spaces of 5-tuples of functions B_1, B_2 is now given by

$$gf(X_1, X_2, X_3) = \begin{pmatrix} B & 0 \\ 0 & I \end{pmatrix} f\,(g^{-1}(X_1, X_2), X_3) \quad , \; f \in B_i, \; g \in E_2$$

Here B is the 2 x 2 matrix which gives the rotation/reflection part of $g \in E_2$ and I is a 3×3 identity matrix. It is a small exercise to check that the Boussinesq equations are indeed equivariant w.r.t. this action. The reason is that the physics is independent of the observer.

In this case it turns out that dim N is infinite. However,
under the assumption that we restrict our attention to solutions which are periodic with respect to a hexagonal lattice,
cf. Figure 31, dim N becomes finite dimensional and an explicit
bifurcation analysis can be carried out [33, 34, cf. also 9].

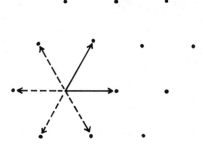

Figure 31

It turns out that the stability of the bifurcation solutions
depends on two parameters in the manner depicted in Figure 32
below. This agrees with experimental data. Of course, this does
not give us a complete mathematical description of the Benard
convection. It remains to be shown that the E_2 symmetry has to
break through a lattice. This is still an open problem, though
it is hard to see how E_2 symmetry could get broken otherwise to
a stable solution (cf. also [38, section II.1]). It also remains to analyse the relative stability of various lattice patterns with respect to each other. Cf. in this connection [31,
32].

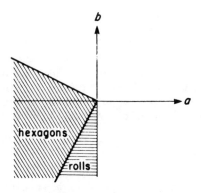

Figure 32

5.5. Drainage patterns. An example . Competing singularities

Let me conclude with a possible model for the formation of

drainage patterns due to Thom [38], which I like particularly.
Imagine a sandy slope on which a gentle rain is continuously
falling. At the top small rivulets will form at random, these
will merge and form larger rivulets further down etc., to form
a pattern somewhat like the one depicted in Figure 33.

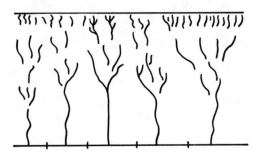

Figure 33

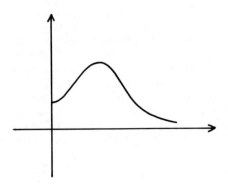

Figure 34

Let X_i denote the position of the i-th watershed near or at the
bottom. Assume that the "erosion power" of a stream is propor-
tial to the width of its basin. Then the position of the i-th
watershed will move according to the differential equation

$$\dot{X}_i = C(X_i - X_{i-1}) - C(X_{i+1} - X_i) = C(2X_i - X_{i-1} - X_{i+1})$$

(5.1)

Obviously equipartition $(X_i = \frac{1}{2}(X_{i-1} + X_{i+1}))$ is a stationary
solution. For an analysis of the stability of such a solution
consider two streams at positions X and -X on R with the divide
at u near zero. Then assuming that C also depends on X we find

$$\dot{u} = 2\,C(X)\,u + 2\,XC'(X)\,u + u^2\,(\ldots\ldots)$$

for small u. So that we will have stability if $C(X) + XC'(X) <$
0. Assuming that $C(X)$ is somewhat like Figure 34, which is not
unreasonable, one would expect a characteristic wave length for
the pattern at bottom given by L = "smallest" X such that $C(X)$
$+ X C'(X) < 0$. The picture of Death Valley some pages back
(Figure 4) is a nice example of just such a pattern with, apparently, a characteristic wave length. It remains of course to
do a complete (bifurcation) analysis which leads from the random pattern at the top to the regular pattern at the bottom.
The governing equation above is a discretisation of a certain
(anti-) diffusion equation. Numerical experiments, [16], with
such equations are encouraging but it seems that we are quite
far from a completely satisfactory theory at the moment.

REFERENCES

1. Bardos, A., Bessis, D. (eds.): 1980, Bifurcation phenomena in mathematical physics and related topics, Reidel Publ. Cy.

2. Bell, TH.L., Wilson, K.G.: 1974, Nonlinear renormalisation groups, Physical Review B 10, 9, pp. 3935-3944.

3. Bensoussan A., Lions, J.L., Papanicolaou, G.: 1978, Asymptotic analysis for periodic structures, North Holland.

4. Brown, G.H, Wolken, J.J.: 1979, Liquid crystals and biological structures, Acad. Press.

5. Chandrasekhar, S.: 1961, Hydrodynamic and hydromagnetic stability, Clarendon Press, Oxford (Reprinted by Dover, 1981).

6. Dixon, R.: The mathematical daisy, New Scientist, 17 December 1981, pp. 792-795.

7. Gardner, M.: Mathematical games: On tesselating the plane with convex polygon tiles, Scientific American, July 1975, pp. 112-117.

8. Gardner, M.: Mathematical games: Extraordinary non periodic tiling that enriches the theory of tiles, Scientific American, January 1977, pp. 110-121.

9. Golubitsky, M.: 1983, The Bénard problem, symmetry and the lattice of isotropy subgroups, In: C.P. Bruter, A. Aragnal, A. Lichnerowicz (eds.), Bifurcation theory, mechanics and physics: Mathematical developments and applications, Reidel Publ. Cy., pp. 225-256.

10. Gonzalez Gascon, F.: 1977, Notes on the symmetries of systems of differential equations, J. Math. Physics 18,9 pp. 1763-1767.

11. Gurel, O., Rössler, O.E. (eds.): 1979, Bifurcation theory and applications in scientific disciplines, Ann. New York Acad. Sci., 316.

12. Haken, W.: 1977, Synergetics, An introduction, Springer.

13. Haken, W.: Synergetics and bifurcation theory, In [11], pp. 357-375.

14. Hazewinkel, M.: April 1978, Letter to J.S. Birman.

15. Hazewinkel, M.: Bifurcation phenomena, A short introductory tutorial with examples, this volume.

16. Hazewinkel, M., Kaashoek, J.F.: Report of the Economic Inst., Erasmus University Rotterdam, in preparation.

17. Kleman, M.: 1983, Geometrical aspects in the theory of defects, In: C.P. Bruter, A. Aragnal, A. Lichnerowicz (eds.), Bifurcation theory, mechanics and physics: mathematical developments and applications, Reidel Publishing Company, pp. 331-356.

18. Marsden, J., Mc Cracken, M.: 1976, The Hopf bifurcation and its applications, Springer.

19. Maurin, K., Raczka, R. (eds.): 1976, Mathematical physics and physical mathematics, Reidel Publ. Cy./PWN.

20. Michel, L.: 1980, Symmetry defects and broken symmetry, Configurations. Hidden symmetry, Reviews of Modern Physics, 52, 3, pp. 617-651.

21. Michel, L.: Simple mathematical models of symmetry breaking, Application to particle physics, In [19], pp. 251-262.

22. Michel, L.: 1970, Applications of group theory to quantum physics, Algebraic aspects, Lect. Notes Physics 6, Springer, pp. 36-143.

23. Prigogine, I.: 1980, From being to becoming, Freeman.

24. Poénaru, V.: 1976, Singularités C^{∞} en présence de symmétrie, Lect. Notes in Math. 510, Springer.

25. Poston, T., Stewart, J.: 1978, Catastrophy theory and its applications, Pitman.

26. Rabinowitz, P.H.: 1977, (ed.), Applications of bifurcation theory, Acad. Pr.

27. Rosen, J.: 1974, Approximate symmetry, Preprint Tel-Aviv Univ.

28. Rosen, J.: 1974, Symmetry and approximate symmetry, Reprint Tel-Aviv University, TAUP-438-74.

29. Ruelle, D.: 1973, Bifurcation in the presence of a symmetry group, Arch. Rat. Mech. Anal. 51, pp. 136-152.

30. Sattinger, D.: 1977, Selection mechanisms for pattern formation, Arch. Rat. Mech. Anal. 66, pp. 31-42.

31. Sattinger, D.: 1978, Group representation theory, bifurcation theory and pattern formation, J. Funct. Anal. 28, pp. 58-101.

32. Sattinger, D.H.: 1973, Topics in stability and bifurcation theory, Lect. Notes in Math. 309, Springer.

33. Sattinger, D.H.: 1980, Bifurcation and symmetry breaking in applied mathematics, Bull. Amer. Math. Soc. 3, pp. 779-819.

34. Sattinger, D.H.: 1979, Group theoretic methods in bifurcation theory, Lect. Notes in Math. 762, Springer.

35. Shelton, J.S.: 1966, Geology illustrated, Freeman.

36. Shubnikov, A.V., Koptsik, V.A.: 1974, Symmetry in science and art, Plenum.

37. Sun, W.Y.: 1982, Cloud bands in the atmosphere, In: E.M. Agee, T. Asai (eds.), Cloud dynamics, Reidel Publ. Cy., pp. 179-194.

38. Thom, R.: Symmetries gained and lost, In [19], pp. 293-320.

39. Thornbury, W.N.: 1969, Principles of geomorphology, Wiley/ Toppan.

40. Waterhouse, W.C.: 1983, Do symmetric problems have symmetric solutions, Amer. Math. Monthly 90, pp. 378-387.

41. Wilson, K.G.: 1976, The renormalization group - Introduction, In: C.S. Domb, M.S. Green, Phase transitions and critical phenomena, Vol. 6, Acad. Press, pp. 1-5.

SYNERGETIC AND RESONANCE ASPECTS OF INTERDISCIPLINARY RESEARCH

Jean H.P. Paelinck[1]
Thijs ten Raa

Erasmus University Rotterdam

"Possibly already stars and the cores of galaxies, but certainly ecosystems, the world-wide Gaia System of the bio-plus atmosphere ..., social systems, civilisations and cultures are no less dissipative self-organizing systems than are ideas, paradigms, the whole system of science, religions and the images we hold of ourselves and of our roles in the evolution of the universe ... ."

> Erich Jantsch in M. Zeleny, Autopoiesis, Dissipative Structures and Spontaneous Social Orders, Westview Press, Boulder, 1980, p. 86

In this essay we shall entertain the thought that the whole system of science has a certain dynamics. The dynamics is made of developments of disciplines intertwined with cross effects, that is, interdisciplinarity. Science does not, however, march down a unique path of development; from certain points on, disciplines may nurture or frustrate each other. Therefore, interdisciplinary research itself may be subjected to bifurcation analysis!

History of science students note that at certain points disciplines merge. For example, at the turn of the century political economy began to employ mathematics and, in return, to breed new chapters of mathematics such as programming, game theory and even some parts of matrix algebra. Two disciplines, traditionally distinct, hooked up; their developments are now

233

M. Hazewinkel et al. (eds.), Bifurcation Analysis, 233–239.
© 1985 by D. Reidel Publishing Company.

simultaneous and may be integrated to some extent. Interdisciplinarity is the buzz word for all this.

The use of mathematics clarifies economic issues and helps
in solving problems of economic theory and policy. Therefore,
mathematics may speed up the development of economics, but
excessive use of mathematics by economists can have choking
effects. The subsequent analysis illustrates this.

Consider two disciplines of science; "indicators", say
$m(t)$ and $e(t)$, which may vary from time to time (t), are
introduced as follows. Each discipline accumulates knowledge;
common parlor for the stock of knowledge is "the state of the
art". The growth rate of the state of the art is, of course, a
(relative) flow variable; it stands for the accumulation of
knowledge in the discipline and is denoted $m(t)$ or $e(t)$. So
these symbols do not refer to the states of both arts or the
sizes of libraries, but to the process of accumulation or
research; $m(t)$ is the research indicator for one discipline,
$e(t)$ for the other.

It is usually assumed that the creation of knowledge is an
irreversible process. Once an invention has been made, it needs
not be redone. Knowledge does not wear of tear; its accumulation rates, $m(t)$ and $e(t)$, are nonnegative. Yet research goes
up and down; its indicators fluctuate. Which forces govern
these fluctuations? In other words, what are the equations
for $\dot{m}(t)$ and $\dot{e}(t)$, where $\cdot$ denotes time differentiation.

It will be fruitless to aim at a single law of research
which holds forever; the "constants" of research, and interdisciplinary research in particular, adjust from time to time.
Instances of such adjustments are the so-called scientific
revolutions: a major new paradigm affects the development of
science. But in between such revolutions research has a dynamics which may be characterized by means of constants; this
dynamics is mathematically described by an autonomous differential equations system with parameters: various stages of
science correspond to different values of the parameters; this,
at least, is one way to model stages of science.

We shall now begin to construct the equations. For focus
we will consider variations of only interdisciplinary parameters; the internal disciplinary parameters are fixed numerically. Specifically, we take

$$\dot{m} = m(1-m), \qquad\qquad\qquad\qquad\qquad\qquad (1)$$

$$\dot{e} = -e. \qquad\qquad\qquad\qquad\qquad\qquad\qquad (2)$$

For either discipline, zero research yields no development.
Otherwise we may divide through by m and e, respectively, to

obtain logarithmic time derivatives and to note the following. The state of the art in the 'm'-discipline has a natural rate of growth equal to one; this is the value of the resonance parameter which, by definition, measures the internal rate of development. If the natural rate is overshot, then 1-m is negative and the discipline slows down; if research is relatively suppressed (m less than one), then 1-m is positive and there is a tendency to catch up. The 'e'-discipline, however, has a hopeless internal dynamics; this discipline, left by its own, contracts its research, until it has exhausted merely a limited amount of the unknown. The value of the resonance parameter or the natural rate of growth of the state of the 'e'-art is zero.

So far we have considered the disciplines in separation, as if they were in the old stage; but now consider the emergence of interdisciplinarity. It seems natural to represent it by the product of the two research indicators, me, the most simple expression in which the indicators reinforce each other. However, the impact on the two disciplines may differ; let it be μme and εme, respectively: μme is inserted in the right hand side of the 'm'-equation, εme is the right hand side of the 'e'-equation. μ and ε are interdisciplinary propensities or synergetic parameters (Greek for "working together"), which, again by definition, measure the interdisciplinarity impact on a unit base.

Although the synergetic parameters will be different, they may be related to each other. Here we have in mind the reluctance of many mathematicians to cooperate with eager economists, out of fear that excessive interdisciplinarity is decremental for mathematical research. Hence μ may be negatively correlated with ε, like μ = 1 - ε. In sum, we postulate the following laws of research:

$$\dot{m} = m(1-m) + (1-\varepsilon)me, \tag{3}$$

$$\dot{e} = -e + \varepsilon me. \tag{4}$$

The laws can be simplified by introducing the logarithmic rate of change operator, $\overset{o}{} = \dfrac{d\log}{dt}$. Then we obtain the Lotka-Volterra equations

$$\overset{o}{m} = 1 - m + (1-\varepsilon)e, \tag{5}$$

$$\overset{o}{e} = \varepsilon m - 1. \tag{6}$$

We will distinguish various cases. ε=0 refers to the old stage
of science in which the e-discipline is distinct from the m-
discipline though the latter receives impulses from the former.
When ε becomes positive we enter the stage of science which is
marked by a moderate degree of interdisciplinarity: 0<ε<1. ε=1
represents a scientific revolution, as we shall see below. Then
ε>1 is entered, the stage marked by an "excessive" flurry of
interdisciplinary activity on the side of the 'e'-discipline.
The first case, ε=0, is similar to the case that has been
discussed already. The other cases are represented by phase
diagrams in Figures 1, 2 and 3.

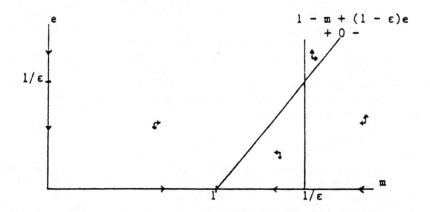

Figure 1. Moderate interdisciplinarity: 0<ε<1.

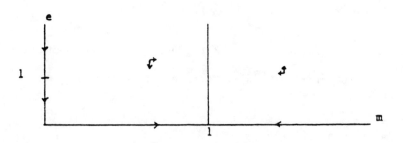

Figure 2. Scientific revolution: ε=1.

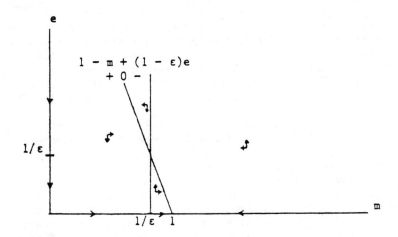

Figure 3. Excessive interdisciplinarity: $\varepsilon>1$.

Throughout, there are three stationary solutions for (m,e), namely $(0,0)$, $(1,0)$ and $(1/\varepsilon,1/\varepsilon)$. In the stage of moderate interdisciplinarity (Figure 1) only $(1,0)$ is stable. In the stage of excessive interdisciplinarity (Figure 3) only $(1/\varepsilon,1/\varepsilon)$ is stable. All these points are plotted in Figures 4 and 5 as functions of ε. The figures also indicate the (in)stability of the points.

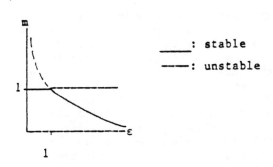

: stable

——: unstable

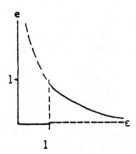

Figure 4.
Stationary m-solutions

Figure 5.
Stationary e-solutions

    We shall briefly discuss the nature of stable stationary
points. For $0<\varepsilon<1$, $(m,e) = (1,0)$ is an attractor; Figure 1
shows that this is true at least for the region where m and e
are both less than $1/\varepsilon$, so the system of science will tend to
the stationary state $(m,e) = (1,0)$. Next consider $\varepsilon>1$. Figure 3
hints that $(1/\varepsilon,1/\varepsilon)$ is the center of counter-clockwise orbits.
In fact, there is a Hopf bifurcation which amounts to a family
of orbits which by the Poincaré-Bendixson theorem spiral down
to $(1/\varepsilon,1/\varepsilon)$ or possibly a limit cycle about it[2]. Starting at
a lower side, interdisciplinarity gives a big push to the 'e'-
discipline; however, this has a slightly decremental effect on
the 'm'-discipline which consequently slows down a bit. This,
on its turn, dries up the source of new results for the 'e'-
discipline which, lacking an own positive force, tumbles down;
the 'm'-discipline is now freed from competitive pressures and
progresses on its own force towards its natural rate of
research. Then the cycle is repeated.
    We have seen in Figures 4 and 5 that at $\varepsilon=1$ two solution
families bifurcate from $(m,e) = (1,1)$. However, both for $\varepsilon<1$
and for $\varepsilon>1$ only one branch was stable; in the latter case this
was $(1/\varepsilon,1/\varepsilon)$. We have also seen that this branch was accompa-
nied by a family of spirals, which is essentially a further
bifurcation. It is a natural question to ask what occurs to
science when, departing from the stage of moderate interdisciplinarity $(0<\varepsilon<1)$ and assuming (m,e) has been attracted already
towards $(1,0)$, the synergetic parameter, ε, is increased to an
"excessive" value greater than one. In the transition there is
the scientific revolution $(\varepsilon=1)$ which is completely unstable as
Figure 2 reveals and even structurally unstable since infinitesimal parametrical change brings about a different picture,
like Figure 1 or 3. Then science may be kicked around anywhere.
Therefore we do not know which orbit will be traced out when
the excessive value of the synergetic parameter is arrived at.
But eventually the system will settle down to a state of either
constant or periodic research as Hirsch and Smale conclude for
a similar case.
    Finally it should be said that, especially in the field
studied, bifurcations are apt to produce themselves in different variations. Two of them have been revealed and discussed in
this essay: multiple (stable) singular parts and structural
instability of the system; the latter would be increased if the
system were to be disturbed by extra non-linearities (apart
from the bilinearity already present), or by hypothesising that
the coefficients could be functions of the state variable. A
third possibility is that, instead of the system of differential equations (5)-(6) one thinks of a system of differential
correspondences, more then one possible timepath originating
from each point (or some points) in the phase plane, but this
possibility is left for further investigation.

Anyway, we have classified various pathes of disciplinary developments which bifurcate in our model of science dynamics, so we have some idea of what the possibilities are. It should be mentioned that this is contingent on our specification of the laws of research; this specification is, of course, just an image we hold of ourselves. And how this image will evolve, bifurcate, ...?

NOTES AND REFERENCES

1) The authors are grateful to Sebastian van Strien for valuable discussions. Netherlands Organization for the Advancement of Pure Research (Z.W.O.) support to the second author is gratefully acknowledged

2) For this we refer to M.W. Hirsch and S. Smale: (1974), _Differential Equations, Dynamical Systems and Linear Algebra_, Academic Press, New York, pp. 264-265.

BIFURCATION: IMPLICATIONS OF THE CONCEPT FOR THE STUDY OF
ORGANISATIONS

Ray Jurkovich

Central Interfaculty
Erasmus University, Rotterdam

1. INTRODUCTION

In this chapter we discuss a few implications of the
bifurcation concept for organisational theory; the notions of
branching and hot point — that point in time and space where a
phenomenon splits — are applied to the structuring processes of
organisations and, to a lesser extent, decision making and
organisational spin-offs (new forms being being established
from older ones). We show that the concept is easy to apply in
at least a few situations in organisational theory; this then
means that the mathematics, which have been presented elsewhere
in this volume, could and should be adopted in organisational
analysis. We begin first with a brief discussion of bifurcation.

2. IDEAL-TYPICAL BIFURCATION

In Hazewinkel's discussion of bifurcations, he sketched a
diagram of what we can label ideal-typical bifurcation where
the process is symmetric (see Figure 1).
This is a description of a phenomenon splitting somewhere
in time and space with one part going off in a direction just
the opposite of the other part and both travelling at the same
rate equidistantly from the dotted line. That point where the
phenomenon splits is called the hot point which can range from
extremely slow (the splitting of rivers) to extremely fast (the
splitting of an atom). In real systems the form of the lines
and amount of symmetry as suggested in Figure 1 are seldom, if
ever, to be found. Rivers, for example, are usually very

M. Hazewinkel et al. (eds.), Bifurcation Analysis, 241–253.
© *1985 by D. Reidel Publishing Company.*

winding and one branch may travel at a considerable distance
from the main trunk relative to another branch; some branches

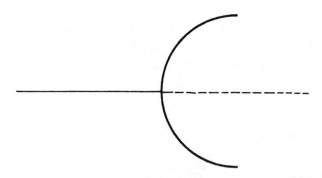

Figure 1. Ideal-typical bifurcation

may also develop more slowly than others. Furthermore, tri-,
quadra-, and even more, "furcations" can develop at the same
point as any casual observation of tree limbs will show: the
entire splitting process can become very complex and its causes
are not always easy to understand. In certain kinds of social
systems (i.e. organisations) certain technological advances
promote their development - their differentiation - which
requires the creation and maintenance of elaborate central components. For example, the development of modern telecommuni-
cations is both the condition and mechanism for the establishment of control, of extremely complex multi-national firms with
subsidiaries scattered around the world[1].

3. BIFURCATION IN ORGANISATIONS

In organisation theory - and especially in the sociology
of organisations - it is quite common to discuss organisation
structure using the hierarchy concept. Some of the newer ideas
such as link-pin and matrix structures do not necessarily mean
that radically different methods of organising have been developed during the past twenty odd years; the notion of hierarchy
to the best of my knowledge has yet to disappear, regardless of
attempts by organisation theorists to develop fancy designs.
Even in chaotic human systems a leader with followers emerges;
establishment of control and the maintenance of its continuity
are critical elements in any organisation; a workable alternative to the hierarchy principle has yet to be developed.

3.1. Vertical and horizontal differentiation (bifurcation)

One of the first steps for both theorists and practitioners when conducting research is to obtain a description of the organisation's structure: this, of course, is enquiry into the nature of the organisation's hierarchy. The image which thus comes to mind is shown in Figure 2.

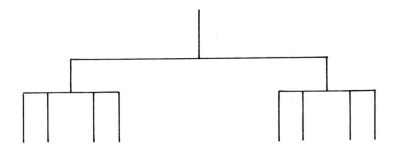

Figure 2. Typical perception of an organisational hierarchy

Now, if we rotate this image to the left (see Figure 3) we then get something which resembles Figure 1.

While it is true that organisation structures in reality differ from that found in formal documents, it is still impossible to deny that people are organised according to the bifurcation principle. While the methods of coordinating may often be different, once again something which is not captured when examining organisation charts, this principle is found in every culture where human cooperation is necessary to pursue goals.

There has been a great deal of research conducted on the development of organisation structures and the implications for system size (belief in an opposite causal relationship is more common). Having said all of this, it is quite obvious that the subject of structural differentiation in organisation theory is what a mathematician interested in bifurcation phenomenon could indeed very easily label structural bifurcation.

Two major types of bifurcation in organised social systems are commonly referred to as vertical differentiation (the number of levels in a hierarchy) and horizontal differentiation (the number of departments per level). A third major type – the division of labour – will be discussed later on in this chapter. The literature is not reviewed here but theories have been generated about the relationships between indicators of differentiation, system size (most often the number of personnel directly involved in production), and the relative size of various overhead components (relative number of supervisors, lower administrative personnel, and other types of personnel

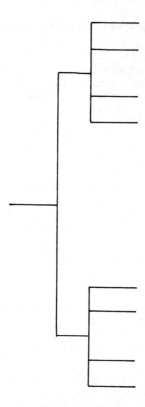

Figure 3. Bifurcation in organisation structure

who provide support or staff services).

Modern theory concerning the nature of the relationships between these things has its origin in the works of Adam Smith, Herbert Spencer, Karl Marx, Emile Durkheim, and, more recently, Pitirim Sorokin[2]. In the 1950's, and especially the 1960's, sociologists for unknown reasons started conducting more studies on this subject; Kimberley has recently reviewed most of the literature[3]. The dominant belief is that increasing system size promotes differentiation of all kinds which in turn means an increase in coordination and communication problems. In order to get these under control more managers, administrative personnel and support or staff personnel are necessary; however, the rate at which they need to be added slows down through time due to the effect of the economy of scale generated as organisations increase in size.

Of special interest here is that bifurcation processes take place during growth while during periods of decline certain branches are either required to decrease their activity or to stop altogether. However, research on the causes and effects of structural differentiaion have led to conflicting results and serious doubts have been casted upon the methodologies used[4]. One major obstacle until recently was that most of the results were obtained from cross-sectional or static research while the bifurcation process in social systems is dynamic; conducting time series analysis of data over long time periods is a possible solution, but the data collection process is very tedious and time consuming while most analysts are under pressure to "publish or perish": the nature of the academic game acts as a brake in setting up and executing research into complicated topics.

In those few cases where time series analysis has been conducted, they have usually been studies over short periods and they often contain "data holes" where data describing certain points in time were not included in the analysis because data were not available. Sometimes a time series analysis, or longitudinal analysis as it is also called in the social sciences, means as little as conducting linear correlation analysis using the difference scores obtainded by subtracting values at T_0 from T_1[5]. This appears to have its problems: regression towards the mean causes deflating correlation coefficients when difference scores over short periods are entered into analysis[6]. In simple cases, where difference scores from a two panel data set over a short period are used in the analysis, we question this approach as an adequate test of the theory. It may be intellectually more rewarding to follow carefully selected quantitative and qualitative aspects of long life cycles of a few organisations as opposed to only a complex of quantitative aspects of a large sample of organisations. The latter approach dominated research very strongly for almost twenty years; this tradition was destabilised with the publication of Organization Life-Cycle in which several essays appear which clearly demonstrate the fruitfulness of the former approach[7]. After reading this set of essays it is then bewildering to look back at all those studies where the results from linear multiple regression analysis of large amounts of data were expected to provide profound insights into organisational growth and development processes. Sadly enough, looking at the world as if it were organised neatly through linear and rank order correlation matrices has hardly helped us gain an understanding of processes which are in reality often bifurcating and dynamic[8].

Why have we allowed ourselves to become channelled into using only certain techniques, techniques which were invented in the previous century? This has had an enormous influence on our thinking. Other theories of nature contain certain ideas which for one reason or another have not been looked at and examined for their utility in other disciplines. For example, Darwin's theory of evaluation is very clearly a bifurcation theory; even though the theory may not in its entity be relevant for the development of all aspects of organisational behaviour, the major principle has very strong implications for constructing descriptive theories.

3.2. The division of labour

3.2.1. Inter-occupational bifurcation. According to Gerard Mensch the development of the division of labour is very clearly a bifurcation process[9]. His theory of this process is provided in Figure 4.

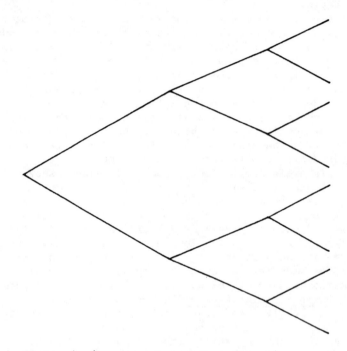

Figure 4. (Figure adapted from Mensch)

Mensch states that new occupations split off from others in the course of time and as new occupations develop some die out. During Man's early history, all people were involved in berry and plant collection. With the development of new tools and organisation a very complex set of occupations has been created. When we look at the development of disciplines within universities we see that up to the Renaissance university academics were exclusively philosophers; the only thing left over from this tradition is the title Ph.D. and departments of philosophy are now one of the smallest within the university system. Nevertheless, all academic disciplines originated from philosophy. Furthermore, centuries ago clergymen were also physicians, but these two occupations nowadays have almost nothing in common; the physical and metaphysical aspects have been nearly completely separated.

Sociologists have devoted a great deal of thought and research on the development of the division of labour; the topic of social stratification – where occupations are the central concept – appears to be one of the most intensively studied in the discipline. Only in the last decade, however, have we seen any major theoretical developments. Giddens and Vanfossen, for example, in departing from the structural-functional paradigm in sociology describe very clearly how occupational groups continuously compete and conflict with each other or, in coalition with each other, struggle against management for scarce resources; they maintain that the development of the division of labour is a dialectical dynamic process[10].

3.2.2. Intra-occupational bifurcation. Furthermore, bifurcation also occurs within occupations; once the existence of an occupation becomes legitimised, different specialities or subdivisions develop through time. When going through the want ads for computer programmers it quickly becomces obvious that as far as programmers are concerned some apparently specialise in certain languages and there is a hierarchy of proficiency (e.g. senior programmer versus junior programmer). This is true for just about any occupational category and probably even more true for certain professions where severe entry and promotion criteria are applied for the purpose of preserving quality which, in turn, enhances their status and justifies their high salaries in the outside world. A good example of an intra-division of labour process is the medical profession: in primitive societies, the medicine man still exists while in industrial society there are many medical specialities. There is a division between general practitioner, medical specialist and medical researcher which is sometimes a combination of the two. Within the field of medical specialities a division can be made between, say, surgeon and non-surgeon, and in both cases there are several types.

4. SUMMARY

Thus, as far as the structuring processes of organisations and the development of the division of labour are concerned, these are quite clearly dynamic bifurcation processes. Organisation theorists, however, have rarely, if ever, analysed these phenomena accordingly; therefore, the dominant techniques are certainly in need of a major overhaul. If this took place then it would be easier to integrate quantitative and qualitative research. As it now stands it is an "either-or" situation. Yet it is precisely the latter approach – qualitative studies of organisational growth and developement –·which has implications for adopting the appropriate mathematical-statistical procedures. Because the results obtained in the past were often dis-

appointing is not necessarily a legitimate reason to wring a
tradition's neck.

5. TWO ADDITIONAL EXAMPLES OF BIFURCATION PROCESSES IN ORGANI-
   SATIONS

There are many processes in organisations which closely
resemble bifurcation processes found in natural systems. We
will only briefly discuss two additional ones in the remainder
of this chapter. These are: 1. decision making, and 2. spin-
offs or the creation of new organisations from existing ones.

5.1. Decision-making processes

The use of decision trees in management policy making is
well-known to every student of organisations. Going through
each step is essentially running through the absence or
presence of certain necessary conditions and events (assuming
for a moment that a policy maker actually knows ahead of time
which conditions are to be taken up in the process) which
steers the selection of an appropriate decision from a known
set of alternatives. This is a neat textbook approach while in
practice it is a lot sloppier; reality seems to range from the
garbage can model to quasi-rational muddling through where subjective and objective information (and their changes) are continuously being processed. But the main thing is that the bifurcation principle seems to be operating whatever the type of
decision making and however rational or irrational it may be.
Understanding it is really quite easy: the human mind does
not process complex problems in one big bite; pieces of the unknown puzzle are added one at a time and after each step a
decision is taken to gather certain types and amounts of information: the cycle continues until enough information has been
obtained to provide the decision maker with a good idea fo what
the solution is. (He never gathers all the information to fully
complete the puzzle since this can be too costly). This amounts
to very complex multi-dimensional "if-then" and "then-select"
statements, or nodes, that have branches leading to other nodes
and, finally, to some end point. Two major problems found in
science are also found in decision making: the validity and
reliability of information; if these are not carefully taken
into consideration during the decision making process then it
is rather obvious that over the short or long run expensive
mistakes can take place. Getting hung out on a limb can turn
out to be a very costly and complex problem. A good example of
this is the decision to invest in nuclear energy. The optimis-
tic statements made by the "experts" in the 1950's regarding
even more and even cheaper electricity influenced decisions to
invest billions in the construction of nuclear plants while

certain safety problems and especially the waste problem have
yet to be solved. A major problem with the process is that if
it ever gets out of hand it happens so fast that human thinking
and action and fail-safe systems are much too slow to be able
to intervene while the consequences are just staggering[11].

5.2. Spin-offs

Our last example of how bifurcation is applicable to organisational behaviour concerns new firm spin-off from older and
larger firms. Nowadays the establishment of new innovative
firms is being encouraged by governments as a partial solution
to a declining economic situation and unemployment. The theory
is that from retrospective analysis it seems that relatively
more product innovation takes place during a preiod of economic
malaise; furthermore, this usually takes place in new, small
and medium size firms and not necessarily in large firms. Encouraging new firm starts and innovation in existing firms is,
of course, a gamble since we really do not know if a forced
repeat of one part of economic history will have the same
effects under very different conditions. Nevertheless, belief
in the theory is very strong: a large amount of venture capital
is now available. There are many accounts of successful new
firms - especially in the electronics industry - in which new
products and processes have been created and where employment
has been generated. Most of these new ventures were established
by entrepreneurs who left firms in which, for various reasons,
there was little or no room for the implementation of their
ideas[12].

The important thing for organisational theory is that when
the general socio-economic environment becomes unfavourable
(e.g. increasing inflation, large government budget deficits,
increasing unemployment and various combinations of these) this
promotes the creation of new organisations. That is, as
unpleasant as it may sound, the threat of economic collapse is
a force which promotes small-scale creativity intended to
increase the chances of survival[13].

There are various types of spin-offs (e.g. a slow grower
versus a fast grower)[14]; they are often eventually taken up in
larger firms (buy-outs); and spin-offs from spin-offs also
occur. The location of spin-offs, or their movement in space,
can be effected by factors in the environment. For example,
during recent interviews with entrepreneurs in the southeastern
corner of the Netherlands, cases were cited of spin-offs locating over the border in West Germany and Belgium. Corporate
profits are taxed less in these two nations (±26% versus 46% in
the Netherlands) and thus the tax climate in the environment
works like a magnet: they attract organisational starts away
from their expected place of establishment.

6. HOT POINTS

This brings us to a discussion of the point in time and
space where behaviour in organised human systems bifurcate.
Under which conditions do occupations in organisations and
organisation structures bifurcate? Under which conditions do
spin-offs take place? The answer to the first question is not
an easy one; in spite of the large body of research on this
topic, we can still only speculate. Obviously the availability
of resources in an organisation's environment plays a critical
role; if resources in the environment are non-existent, then
bifurcation will not take place. Environmental demand, in our
opinion, provides the only legitimate reason for the start and
development of organisational complexity. Demand can, of
course, be stimulated but there are limits and the consequences
can approach disaster. For example, deficit financing of
governmental budgets worked when economies were growing but now
that the economic situation has taken a turn in the opposite
direction there is a tendency to abolish departments, merge
existing ones, and freeze or cut civil service salaries. In the
private sector the situation is more complex: large rigid business bureaucracies with declining profits and losses have fired
thousands of personnel (e.g. General Motors and Ford) while
high technology firms which have discovered flourishing market
niches are expanding.

Answers to the secon question are less difficult to find
in spite of the newness of this research tradition. Sudden
drops in taxes on small firm corporate profits had an almost
overnight effect on the creation of new firms in the United
States. There are countless stories of dissatisfied young
engineers leaving large firms to develop new products which top
management felt were not in the interest of preserving the
organisation's continuity. There are also socio-cultural
factors which influence new starts. From results obtained
during the first phase of research into problems and possibilities that small and medium size firms in the electro-technical
industry in the Netherlands are experiencing, it appears that
going into business for yourself is regarded as a very risky
business and there is a fear to fail. As a consequence there is
more emphasis on career stability and security in the Netherlands than, for example, in the United States. Mobility and
risk are regarded as something to avoid. How this is embedded
in the social system was reflected in the statement of one
entrepreneur: "The Dutch are just not inclined to do that kind
of thing." The financing system also plays a role: it was
repeatedly stated that the banks are extremely wary of providing loans for new starts which might take two to three years
before a break-even point is reached. Finally, a complex network of national and local legal constraints has discouraged
new starts; only recently has there been discussion of "deregu-

lation" but in practice until now very little has been done.

7. SUMMARY AND CONCLUSIONS

It is quite obvious that the bifurcation concept does indeed have implications for organisational analysis. Only a few examples were touched upon in this paper but with a further search it would not surprise us if the list would provide sufficient justification for incorporating the concept in future research. In doing so it also implies adopting mathematical-statistical procedures which are suited to the reality being studied. Inferring dynamic hypotheses from static data has led to more confusion than understanding of changing man-made authority structures and patterned behaviour; it is no wonder then that linear analysis of non-linear systems has resulted in a "cemetry of hypotheses" and not even a good start in the direction of a grand theory of organisations.

Now that organisation theory has entered a period of a healthy interaction between theorists and practice, it is also time to look at the tools used in other disciplines for the very simple reason that they contain ideas which can help us gain a profound understanding of reality. There is, of course, nothing new about this; most contemporary organisation theorists discuss organisations as systems and this is evidence of the influence of system theory which took place in the 1960's. A good example of how biosystem thinking is creatively used can be found in Aldrich's treatment of the organisation-environment relationship; the general idea has entered and has been accepted although the concepts are not the same.

Finally, we are not saying that all mathematical-statistical tools should be incorporated in organisation theory. What we are saying is that there are certain processes in organised human systems which are often, but not always, very similar to those found in natural systems for which analytical tools are available that in turn can be used to study organisational behaviour.

NOTES AND REFERENCES

1. While multinational firms existed before the invention of
   telegraph and telephone, the ease of establishing, and
   especially controlling the behaviour of subsidiaries is
   much easier nowadays.

2. Smith, Adam: 1904 (1776), An Inquiry Into the Nature and
   Causes of the Wealth of Nations, Edwin Cannon (ed.),
   Methuen, London;
   Spencer, Herbert: 1877, The Principle of Sociology, Vol. 1,
   D. Appleton and Company, New York;
   Durkheim, Emile: 1964 (1893), The Division of Labour in
   Society, Free Press, New York;
   Sorokin, Pitirim A.: 1964 (1941), Social and Cultural Mobi-
   lity, The Free Press of Glencoe, Collier-Macmillan Limited,
   London.

3. Kimberley, John R.: Organizational Size and the Structuralist Perspective: a review, critique, and proposal, Administrative Science Quarterly, 21, pp. 571-597.

4. Kimberley, ibid.: John Freeman and J.E. Kronenfeld,
   Problems of definitional dependency: the case of administrative intensity", Social Forces, 52, pp. 108-121;
   Dogramaci, Ali.: Research on the size of administrative
   overhead and productivity: some methodological considerations, Administrative Science Quarterly, 22, pp. 2-29;
   Jurkovich, R.: 1982, Administrative Structure and Growth: A
   Longitudinal Analysis, Ph.D. thesis, unpublished, Rotterdam.

5. See, for example, Gary E. Hendershot and T.F. James: Size
   and growth as determinants of administrative production
   ratios in organizations, American Sociological Review, 37,
   pp. 149-153.

6. Pendelton, Brian F., Warren R.D. and Chang, H.C.: Correlated denominators in multiple regression and change
   analysis, Sociological Methods and Research, 7/4, pp. 451-
   474.

7. Kimberley, John and Miles, Robert: 1980, The Organizational
   Life Cycle, Jossey-Bass Publishers, London.

8. It should be emphasized that just because this volume is
   concerned with bifurcation does not mean that all phenomena
   in social systems are reflect bifurcation processes. Our
   contention is, however, that bifurcation has been neglected.

9. Mensch, Gerhard: 1979, Stalemate in Technology, Cambridge, Cambridge Mass.

10. Giddens, A.: 1977, The Class Structure of Advanced Societies, London;
    Vanfossen, B.E.: 1979, The Structure of Social Inequality, Little, Brown and Co., Boston.

11. Laporte, Todd: 1982, On the Design and Management of Nearly Error-Free Organizational Control Systems, in: D.L. Sills, C.P. Wolf and V.B. Shelanski (eds.) Accident at Three Mile Island: the Human Dimensions, Westview Press Inc., Boulder Co., pp. 185-200.

12. See, for example, Shapero, Albert: 1980, The entrepreneur, the small firm and possible policies, Six Countries Programme Workshop on Entrepreneurship, Limerick, Ireland.

13. Rothwell, Roy and Zegveld, Walter: 1982, Innovation and the Small and Medium Sized Firm, Francis Pinter, London; Mensch, op. cit.

14. The concepts slow and fast growth are taken from John Freeman and Michael Hannan: Growth and decline processes in organizations, American Sociological Review, 40, pp. 215-228.

INDEX